TI-84 Plus Graphing Calculator

FOR DUMMIES®

A Wiley Brand

2nd Edition

TI-84 Plus
Graphing Calculator

FOR
DUMMIES®
A Wiley Brand

2nd Edition

by Jeff McCalla and C.C. Edwards

FOR
DUMMIES®
A Wiley Brand

TI-84 Plus Graphing Calculator For Dummies® 2nd Edition

Published by
John Wiley & Sons, Inc.
111 River Street
Hoboken, NJ 07030-5774

www.wiley.com

About the Authors

Jeff McCalla is currently teaching mathematics and coaching the golf team at St. Mary's Episcopal School in Memphis, TN. Jeff holds a Bachelor's Degree in Christian Education with a minor in mathematics from Wheaton College and a Master of Arts in Teaching degree from the University of Memphis. Jeff is the co-founder of the TI-Nspire SuperUser group, dedicated to providing advanced training from the world's foremost experts. In addition, he enjoys traveling the country, training teachers as an instructor for Texas Instruments T^3 Program. Jeff also wrote *TI-Nspire For Dummies*. A highlight for Jeff was receiving the Presidential Award for Excellence in Science & Mathematics Teaching and with it, the opportunity to meet both President Obama and Secretary of Education, Arne Duncan. When he is not meeting important dignitaries, Jeff enjoys going to ballgames with his boys, playing golf and tennis, eating Chick-fil-A, reading Malcolm Gladwell and John Wooden, getting free stuff, teaching Sunday school, and making his wife smile.

C. C. Edwards has a Ph.D. in mathematics from the University of Wisconsin, Milwaukee, and is currently teaching mathematics on the undergraduate and graduate levels. She has been using technology in the classroom since before Texas Instruments came out with their first graphing calculator, and she frequently gives workshops at national and international conferences on using technology in the classroom. She has written forty activities for Texas Instrument's Explorations web site, and she was an editor of *Eightysomething*, a newsletter formerly published by Texas Instruments. She is also the author of *TI-83 Plus Graphing Calculator For Dummies*.

Just barely five feet tall, CC, as her friends call her, has three goals in life: to be six inches taller, to have naturally curly hair, and to be independently wealthy. As yet, she is nowhere close to meeting any of these goals. When she retires, she plans to become an old lady carpenter.

Dedication

This book is dedicated to my parents, Bud & Elaine McCalla, whose passion for mathematics rubbed off on me. I am most thankful for them not tossing me to the curb when I went through my annoying middle school years.

Author's Acknowledgments

I could not have written this book without the help and support of the people at Wiley. First and foremost, I want to thank my project editor, Blair Pottenger, whose expertise was indispensable in the writing process. I also want to thank my acquisitions editor, Amy Fandrei, for keeping me to a tight schedule and helping me with the content and outline of the book. It is my pleasure to thank Debbye Butler for her thoroughness and copy edit. Additionally, I want to thank Fred Decovsky, Ed. D., for verifying the mathematical and technical accuracy of this book.

I certainly want to thank my friends at Texas Instruments for their ongoing support. The leadership of Gayle Mujica, Charlyne Young, and Kevin Spry has helped make this project possible. In particular, Margo Mankus has come through for me every time I had questions about the new features on the TI-84 C. And, Lydia Neher for always keeping me up to date on the most recent OS changes. Of course, I want to thank Tonya Hancock for her role in fueling my calculator addiction.

Fellow T[3] instructors have assisted me with their help when called upon. Jennifer Wilson's help has been instrumental in the writing of this book. Her eye for detail and general grammar mastery continue to amaze me. I want to thank Jill Gough, who constantly challenges my thinking about teaching. Bryson Perry's technical expertise has been an invaluable resource.

On the home front, I wish to thank my teaching colleagues, Orion Miller, Sandra Halfacre, and Chrystal Hogan, who were nice resources when I had questions. I need to thank the administration at my school, Albert Throckmorton and Patti Ray, for their leadership and for allowing me to pursue my writing. I also want to thank my wife Shannon and my three boys, Matt, Josh, and Caleb, for putting up with me during the writing process. Finally, I want to thank the students I teach at St. Mary's Episcopal School, who are the inspiration for much of what I do.

Publisher's Acknowledgments

We're proud of this book; please send us your comments at http://dummies.custhelp.com. For other comments, please contact our Customer Care Department within the U.S. at 877-762-2974, outside the U.S. at 317-572-3993, or fax 317-572-4002.

Some of the people who helped bring this book to market include the following:

Acquisitions and Editorial

Project Editor: Blair J. Pottenger

Acquisitions Editor: Amy Fandrei

Copy Editor: Debbye Butler

Technical Editor: Fred Decovsky, Ed. D.

Editorial Manager: Kevin Kirschner

Editorial Assistant: Annie Sullivan

Sr. Editorial Assistant: Cherie Case

Cover Photo: Background: © iStockphoto.com/ Andrew Rich; Calculator: Wiley

Composition Services

Senior Project Coordinator: Kristie Rees

Layout and Graphics: Joyce Haughey, Ron Wise

Proofreaders: John Greenough, Evelyn Wellborn

Indexer: BIM Indexing & Proofreading Services

Publishing and Editorial for Technology Dummies

Richard Swadley, Vice President and Executive Group Publisher

Andy Cummings, Vice President and Publisher

Mary Bednarek, Executive Acquisitions Director

Mary C. Corder, Editorial Director

Publishing for Consumer Dummies

Kathleen Nebenhaus, Vice President and Executive Publisher

Composition Services

Debbie Stailey, Director of Composition Services

Contents at a Glance

Table of Contents

Introduction

● ●

Do you know how to use the TI-84 Plus or TI-84 Plus C family of calcula-tors to do each of the following?

✔ Access hidden shortcut menus

✔ Graph functions, inequalities, or transformations of functions

✔ Copy and paste expressions

✔ Insert an image as the background of a graph (TI-84 Plus C)

✔ Write calculator programs

✔ Transfer files between two or more calculators

✔ Create stat plots and analyze statistical data

✔ Graph scatter plots, parametric equations, polar equations, and sequences

If not, then this is the book for you. Contained within these pages are straightforward, easy-to-follow directions that show you how to do every-thing listed here — and much, much more.

About This Book

The TI-84 Plus calculator is capable of doing a lot of things, and this book shows you how to utilize its full potential.

It covers more than just the basics of using the calculator, paying special attention to warn you of the problems that you could encounter if you know only the basics of using the calculator.

This is a reference book. It's process-driven, not application-driven. You won't be given a problem to solve and then be told how to use the calculator to solve that particular problem. Instead, you're given the steps needed to get the calculator to perform a particular task, such as constructing a histo-gram or graphing a scatter plot.

Conventions Used in This Book

When I refer to "your calculator," I am referring to the TI-84 Plus and TI-84 Plus C family of calculators because the keystrokes on these calculators are almost the same. When I want you to press a key on the calculator, I use an icon for that key. For example, if I want you to press the ENTER key, I say press ENTER. If I want you to press a series of keys, such as the Stat key and then the right-arrow key, I say (for example) press STAT ▶. All keys on the calculator are pressed one at a time — there is no such thing as holding down one key while you press another key.

It's tricky enough to get familiar with the location of the keys on the calculator, and even more of a challenge to remember the location of the secondary functions, such as the blue functions that appear above the key. So when I want you to access one of those functions, I give you the actual keystrokes. For example, if I want you to access the Angle menu, I tell you to press 2nd APPS. This is a simpler method than that of the manual that came with your calculator — which would say press 2nd [ANGLE] and then make you hunt for the location of the secondary function ANGLE. The same principle holds for using key combinations to enter specific characters; for example, I tell you to press ALPHA 0 to enter a space.

When I want you to use the arrow keys, but not in any specific order, I say press the ▶ ◀ ▲ ▼ keys or use the arrow keys. If I want you to use only the up- and down-arrow keys, I say press ▲ ▼.

All of the screenshots in this book were taken using the TI-84 Plus C calculator. Of course, you will only be able to see color screenshots in the color insert pages.

What You're Not to Read

The items that follow a Technical Stuff icon are designed for the curious reader who wants to know — but doesn't really need to know — why something happens.

Sidebars provide optional reading that you may find interesting. Feel free to skip reading the sidebars if you want since they will not contain crucial information related to your understanding of the topic. The sidebars are strategically placed, 'extras' that could only enhance your learning.

Foolish Assumptions

My nonfoolish assumption is that you know (in effect) nothing about using the calculator, or you wouldn't be reading this book. My foolish assumptions are as follows:

✔ You own, or have access to, one of the TI-84 Plus or TI-84 Plus C family of calculators.

✔ If you want to transfer files between your calculator and your computer, I assume that you have a computer and know the basics of how to operate it.

How This Book Is Organized

The parts of this book are organized by tasks that you would like to have the calculator perform.

Part I: Making Friends with the Calculator

This part describes the basics of using the calculator. It addresses such tasks as adjusting the contrast and getting the calculator to perform basic arithmetic operations. It also explains how to deal with fractions and how to solve equations.

Part II: Taking Your Calculator Relationship to the Next Level

This part shows you how to enter and evaluate complex numbers. It also introduces you to some of the most useful menus for solving problems. Everything from the basics of converting fractions and decimals to entering and storing matrices is covered in this part.

Part III: Graphing and Analyzing Functions

In this part, think visual. Part III shows you how to graph and analyze functions, inequalities, and transformations of functions. It even explains how to create a table for the graph, inequality, or transformation. In addition, you learn to graph parametric equations, polar equations, and sequences.

Part IV: Working with Probability and Statistics

It's highly probable that Part IV will show you not only how to deal with probability and statistics, but also how to enter data in lists and perform regressions. Learn to seed a random number in your calculator so you can amaze your friends by predicting randomly generated numbers. As a bonus, you learn to use the binomial theorem to expand expressions.

Part V: Doing More with Your Calculator

Part V describes how you can save calculator files on a computer and how you can transfer files from one calculator to another. Find out how to use TI-Connect software to transfer a color photo image from the computer to a graph on your calculator. This part also shows you how to use the Finance app to make the power of compound interest work for you. You also learn to archive and group files in order to manage the memory on your calculator and avoid common errors.

Part VI: The Part of Tens

Part VI contains a ton of useful information packaged nicely in groups of ten. Learn to use the essential skills that you need to succeed in the classroom. This part also describes the most common errors and error messages that you may encounter.

Icons Used in This Book

This book uses four icons to help you along the way. Here's what they are and what they mean:

The text following this icon tells you about shortcuts and other ways of enhancing your use of the calculator.

The text following this icon tells you something you should remember because if you don't, it may cause you problems later. Usually the Remember icon highlights a reminder to enter the appropriate type of number so you can avoid an error message.

There is no such thing as crashing the calculator. But this icon warns you of those *few* times when you can do something wrong on the calculator and be totally baffled because the calculator is giving you confusing feedback — either no error message or a cryptic error message that doesn't really tell you the true location of the problem.

This is the stuff you don't need to read unless you're really curious.

Where to Go from Here

This book is designed so you do not have to read it from cover to cover. You don't even have to start reading at the beginning of a chapter. When you want to know how to get the calculator to do something, just start reading at the beginning of the appropriate section. The Index and Table of Contents should help you find whatever you're looking for. And for your first tip on where to find other information:

You can find an additional topic — doing geometry using Cabri Jr. — on the official Dummies website. To check it out, visit www.dummies.com/go/ti84 and then click on the Downloads tab.

Part I
Making Friends with the Calculator

getting started with

TI-84 Plus

In this part . . .

✔ Get familiar with the basics of your calculator — from turning it on and using menus to changing the mode and accessing the catalog.

✔ Find out how to enter and evaluate expressions, store variables, and work in scientific notation.

✔ Learn to access the fraction templates in the shortcut menu as well as how to convert decimals to fractions.

✔ See how to use the equation solver to make strategic guesses in order to solve equations.

✔ Discover how you can use the PlySmlt2 app to find the roots of a polynomial and solve a system of equations.

Chapter 1

Starting with the Basics

. . . .

In This Chapter

▶ Turning the calculator on and off

▶ Using the keyboard

▶ Utilizing the menus

▶ Setting the mode of the calculator

▶ Using the Catalog

. . . .

*T*he most popular calculator in the world just got a makeover! In this book, you find out how to take advantage of the improvements that have been made to the TI-84 Plus, as well as all of the built-in functionality that has not changed. The best way to use your calculator to the fullest is to read this book and start playing with the device.

The TI-84 Plus C Silver Edition graphing calculator is loaded with many useful features. With it, you can solve equations of all types. You can graph and investigate functions, parametric equations, polar equations, and sequences. You can use it to analyze statistical data and to manipulate matrices. You can even use it to calculate mortgage payments.

What if you own the TI-84 Plus and not the TI-84 Plus C? No worries! The vast majority of the steps will be exactly the same for both calculators. You'll see a difference in the appearance of the graph screen — the TI-84 Plus C has a higher resolution color screen. If you own the TI-84 Plus, ignore any steps referencing color and skip Chapter 22 (about inserting color images) altogether.

If you've never used a graphing calculator before, you may at first find it a bit intimidating. After all, it contains about two dozen menus, many of which contain three or four submenus. But it's really not that hard to get used to using the calculator. After you get familiar with what the calculator is capable of doing, finding the menu that houses the command you need is quite easy. And you have this book to help you along the way.

Why Didn't I Think of That?

You may have the same reaction that I did to some of the changes that have been made to the calculator: "Why didn't I think of that?" It's possible that you did actually! Many of the changes to the TI-84 Plus are a direct result of feedback received from teachers and students. After all, Texas Instruments is committed to providing the best tools for the teaching and learning of mathematics and science.

What does the C stand for in TI-84 Plus C? Color! Say goodbye to having trouble distinguishing functions when you're graphing more than one function on the same screen. Although some of the improvements are subtle, you'll notice others the first time you pick up your new calculator. Here's a small sampling of the changes:

- **New menu options:** I love that all additional menu options have been strategically placed at the end of menus. For example, a new option in the Stat CALC menu, QuickPlot & Fit–EQ as illustrated in the first screen in Figure 1-1. Have you memorized keystrokes, like ZOOM 6 for ZStandard? No problem! The functionality you know hasn't changed.

- **Status Bar:** A quick glance at the top of your screen informs you of the mode settings (like Radian or Degree) as well as a battery status icon. See the top of any of the screens in Figure 1-1. The Status Bar is always there whether you're working on the current line of the Home screen, graph, or table!

- **Higher resolution LCD backlit screen:** Not quite HD quality, but the new screen has more than seven times as many graph area pixels as the original (266×166 versus 96×64)! Plus, you can work on problems at night on a screen that's backlit.

- **Border on graph screen:** Helpful info like function names and coordinates of intersection points are kept separate from the graph, as shown in the second screen in Figure 1-1. Whoever thought of this is brilliant!

- **Table enhancements:** Separator lines and color-coded lists (matching the functions) are more pleasing to the eye. Built-in tips called Context Help are located at the top of the screen, including hints like Press + for ΔTbl. Check out the new table look in the third screen in Figure 1-1.

Think you've seen it all? Not even close. I explain these improvements and much more — just keep reading.

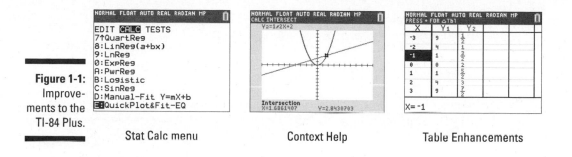

Figure 1-1:
Improvements to the TI-84 Plus.

Stat Calc menu   Context Help   Table Enhancements

Charging the Battery

The TI-84 Plus C uses a Li–ion battery, similar to the one in your cellphone, that holds a charge for up to two weeks. Texas Instruments (TI) recommends charging your battery for at least four hours for peak performance. On the right side of your calculator, an LED light lights up during the recharging process. An amber color indicates your calculator is charging, and a green color indicates your calculator is fully charged. There are three ways to recharge your calculator battery:

The TI-84 Plus does not have a rechargeable battery. You must open the back panel and insert four new AAA batteries.

✔ **TI Wall Adapter:** Simply plug in the adapter that came bundled with your calculator.

✔ **USB computer cable:** Use the USB computer cable that came with your calculator and a computer to charge your calculator. Plug the USB hub into the computer and plug the mini-USB hub into your calculator.

Your computer may not recognize the USB computer cable you are using to charge your calculator. If this happens, download TI-Connect software from http://education.ti.com. For more details on downloading and installing TI-Connect, see Chapter 20.

✔ **TI-84 C Charging Station:** If your classroom has one of these, simply place your calculator in one of the slots of the charging station.

In the top-right part of the screen, a battery status icon indicates the battery level. There are four different battery levels plus a charging icon, as shown in Figure 1-2.

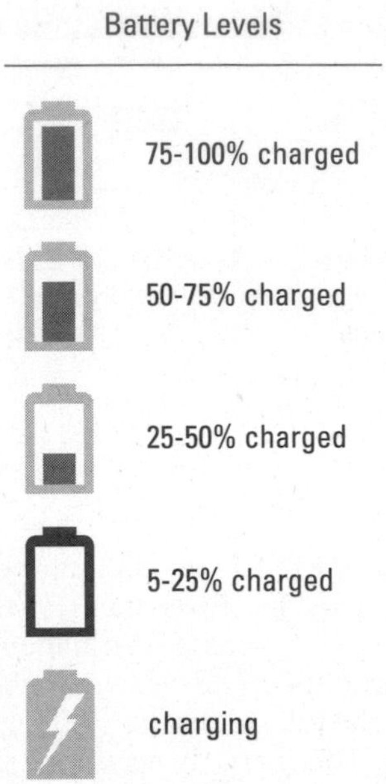

Battery Levels

75-100% charged

50-75% charged

25-50% charged

Figure 1-2:
Battery
status icon
battery
levels.

5-25% charged

charging

WARNING!

If your battery loses its charge, the RAM memory on your calculator may be cleared. If you have programs or data that you don't want to lose, back up your calculator (see Chapter 23 for more details). Your calculator gives you a warning message, as shown in Figure 1-3.

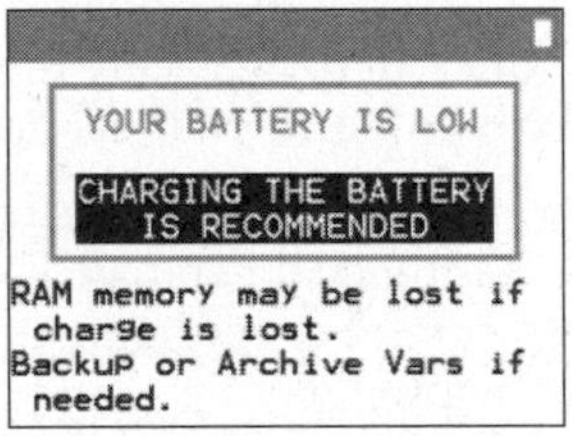

Figure 1-3:
Battery level
warning
screen.

```
YOUR BATTERY IS LOW
CHARGING THE BATTERY
   IS RECOMMENDED
RAM memory may be lost if
  charge is lost.
Backup or Archive Vars if
  needed.
```

Turning the Calculator On and Off

Press [ON] to turn the calculator on. To turn the calculator off, press [2nd] and then press [ON]. These keys are in the left column of the keyboard. The [ON] key is at the bottom of the column, and the [2nd] key is the second key from the top of this column.

To prolong the life of the batteries, the calculator automatically turns itself off after five minutes of inactivity. But don't worry — when you press ⟨ON⟩, all your work will appear on the calculator just as you left it before the calculator turned itself off.

The first time you turn on your calculator, you're greeted by an information screen, as shown in Figure 1-4. A few helpful reminders are displayed on the information screen. If you want to see this screen the next time you turn on your calculator, press ⟨2⟩. Otherwise, press ⟨1⟩ or ⟨ENTER⟩.

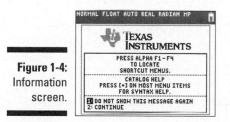

Figure 1-4: Information screen.

In some types of light, the screen can be hard to see. To increase the contrast, press and release ⟨2nd⟩ and then hold down ⟨▲⟩ until you have the desired contrast. To decrease the contrast, press ⟨2nd⟩ and hold ⟨▼⟩.

Using the Keyboard

The row of keys under the calculator screen contains the keys you use when graphing. The next three rows, for the most part, contain editing keys, menu keys, and arrow keys. The arrow keys (⟨▶⟩⟨◀⟩⟨▲⟩⟨▼⟩) control the movement of the cursor. The remaining rows contain, among other things, the keys you typically find on a scientific calculator.

Keys on the calculator are always pressed one at a time; they are *never* pressed simultaneously. In this book, an instruction such as ⟨2nd⟩⟨ON⟩ indicates that you should first press ⟨2nd⟩ and then press ⟨ON⟩.

Accessing the functions in blue

Above and to the left of most keys is a secondary key function written in blue. To access that function, first press ⟨2nd⟩ and then press the key. For example, π is in blue above the ⟨^⟩ key, so to use π in an expression, press ⟨2nd⟩ and then press ⟨^⟩.

Because hunting for the function in blue can be tedious, in this book I use only the actual keystrokes. For example, I make statements like, "π is entered into the calculator by pressing [2nd][^]." Most other books would state, "π is entered into the calculator by pressing [2nd] [π]."

When the [2nd] key is active and the calculator is waiting for you to press the next key, the blinking ■ cursor symbol is replaced with the 🛈 symbol.

Using the [ALPHA] key to write words

Above and to the right of most keys is a letter written in green. To access these letters, first press [ALPHA] and then press the key. For example, because the letter O is in green above the [7] key, to enter this letter, press [ALPHA] and then press [7].

Because hunting for letters on the calculator can be tedious, I tell you the exact keystrokes needed to create them. For example, if I want you to enter the letter O, I say, "Press [ALPHA][7] to enter the letter O." Most other books would say "Press [ALPHA] [O]" and leave it up to you to figure out where that letter is on the calculator.

You must press [ALPHA] before entering each letter. However, if you want to enter many letters, first press [2nd][ALPHA] to lock the calculator in Alpha mode. Then all you have to do is press the keys for the various letters. When you're finished, press [ALPHA] to take the calculator out of Alpha mode. For example, to enter the word TEST into the calculator, press [2nd][ALPHA][4][SIN][LN][4] and then press [ALPHA] to tell the calculator that you're no longer entering letters.

When the calculator is in Alpha mode, the blinking ■ cursor symbol is replaced with the 🅰 symbol. This symbol indicates that the next key you press will insert the green letter above that key. To take the calculator out of Alpha mode, press [ALPHA].

Using the [ENTER] key

The [ENTER] key is used to evaluate expressions and to execute commands. After you have, for example, entered an arithmetic expression (such as 5 + 4), press [ENTER] to evaluate that expression. In this context, the [ENTER] key functions as the equal sign. Entering arithmetic expressions is explained in Chapter 2.

Using the $\boxed{X,T,\Theta,n}$ key

$\boxed{X,T,\Theta,n}$ is the key you use to enter the variable in the definition of a function, a parametric equation, a polar equation, or a sequence. In Function mode, this key produces the variable **X**. In Parametric mode, it produces the variable **T**; and in Polar and Sequence modes, it produces the variables θ and *n*, respectively. For more information, see the "Setting the Mode" section later in this chapter.

Using the arrow keys

The arrow keys ($\boxed{\blacktriangleright}$, $\boxed{\blacktriangleleft}$, $\boxed{\blacktriangle}$, and $\boxed{\blacktriangledown}$) control the movement of the cursor. These keys are in a circular pattern in the upper-right corner of the keyboard. As expected, $\boxed{\blacktriangleright}$ moves the cursor to the right, $\boxed{\blacktriangleleft}$ moves it to the left, and so on. When I want you to use the arrow keys — but not in any specific order — I refer to them all together, as in "Use the $\boxed{\blacktriangleright}\boxed{\blacktriangleleft}\boxed{\blacktriangle}\boxed{\blacktriangledown}$ keys to place the cursor on the entry."

Keys to remember

The following keystroke and keys are invaluable:

- $\boxed{\text{2nd}}\boxed{\text{MODE}}$: This is the equivalent of the Escape key on a computer. It gets you out of whatever you're doing (or have finished doing) and returns you to the Home screen. See the next section for more about the Home screen.

- $\boxed{\text{ENTER}}$: This key is used to execute commands and to evaluate expressions. When evaluating expressions, it's the equivalent of the equal sign.

- $\boxed{\text{CLEAR}}$: This is the "erase" key. If you enter something into the calculator and then change your mind, press this key. If you want to erase the contents of the Home screen, repeatedly press this key until the Home screen is blank.

What Is the Home Screen?

The Home screen is the screen that appears on the calculator when you first turn it on. This is the screen where most of the action takes place as you use the calculator — it's where you evaluate expressions and execute commands. This is also the screen you usually return to after you've completed a task such as entering a matrix in the Matrix editor or entering data in the Stat List editor.

Press [2nd][MODE] to return to the Home screen from any other screen. This combination of keystrokes, [2nd][MODE], is the equivalent of the Escape key on a computer. It always takes you back to the Home screen.

If you want to clear the contents of the Home screen, repeatedly press [CLEAR] until the Home screen is blank.

The Busy Indicator

If you see a moving dotted ellipse in the upper-right corner of the screen, this indicates that the calculator is busy graphing a function, evaluating an expression, or executing a command.

If it's taking too long for the calculator to graph a function, evaluate an expression, or execute a command, and you want to abort the process, press [ON]. If you're then confronted with a menu that asks you to select either **Quit** or **Goto**, select **Quit** to abort the process.

Editing Entries

The calculator offers four ways to edit an entry:

✔ **Deleting the entire entry:**

   Use the [▶][◀][▲][▼] keys to place the cursor anywhere in the entry and then press [CLEAR] to delete the entry.

✔ **Erasing part of an entry:**

   To erase a single character, use the [▶][◀][▲][▼] keys to place the cursor on the character you want to delete and then press [DEL] to delete that character.

✔ **Inserting characters:**

   Because "typing over" is the default mode, to insert characters you must first press [2nd][DEL] to enter Insert mode. When you insert characters, the inserted characters are placed to the left of the cursor. For example, if you want to insert CD between B and E in the word ABEF, you would place the cursor on E to make the insertion.

To insert characters, use the ▶◀▲▼ keys to place the cursor at the location of the desired insertion, press 2nd DEL, and then key in the characters you want to insert. Notice, the cursor does not blink with the typical ■ you're used to seeing; instead, it blinks with an underscore. When you're finished inserting characters, press one of the arrow keys to take the calculator out of Insert mode.

✔ **Keying over existing characters:**

"Type over" is the default mode of the calculator. So if you want to over-type existing characters, just use the ▶◀▲▼ keys to put the cursor where you want to start, and then use the keyboard to enter new characters.

Copying and Pasting

Save time by not retyping similar expressions from scratch! Press 2nd MODE to access the Home screen.

Press the ▲ key to scroll through your previous calculations. When a previous entry or answer is highlighted, press ENTER to paste it into your current entry line. See the first two screens in Figure 1-5.

After you have pasted the expression into the current entry line, you can edit the expression as much as you like. See the third screen in Figure 1-5.

If the answer is in the form of a list or matrix, it cannot be copied and pasted. Instead, copy and paste the expression. Also, notice that the mode settings don't display in the Status bar when you're scrolling through the calculator history.

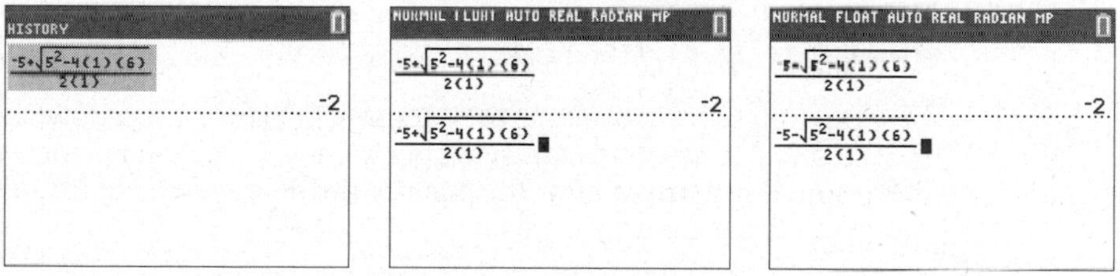

Figure 1-5:
Copying and pasting.

| Highlight expression | Press ENTER | Edited expression |

Using Menus

Most functions and commands that you use are found in the menus housed in the calculator — and just about every chapter in this book refers to them. This section is designed to give you an overview of how to find and select menu items.

Accessing a menu

Each menu has its own key or key combination. For example, to access the Math menu, press MATH; to access the Test menu, press 2nd MATH. An example of a menu appears in the first screen in Figure 1-6. This is a picture of the Math menu.

Some menus, such as the Math menu, contain submenus. This is also illustrated in the first screen in Figure 1-6. This screen shows that the submenus in the Math menu are MATH, NUM, CMPLX, PROB, and FRAC (Math, Number, Complex, Probability, and Fraction). Use the ▶◀ keys to view the items on the other submenus. This is illustrated in the second and third screens in Figure 1-6.

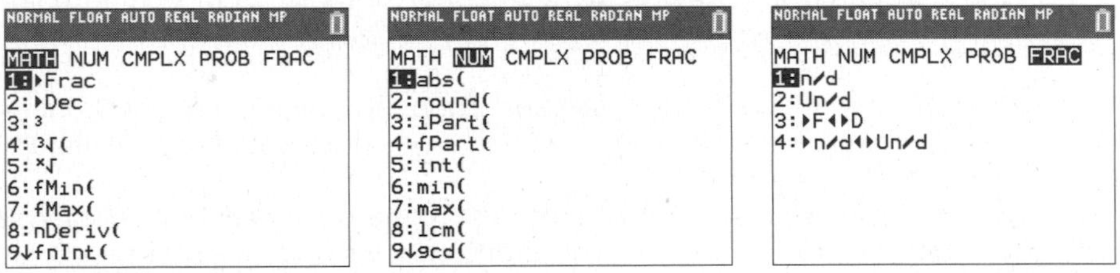

Figure 1-6: Submenus of the Math menu.

Math MATH menu   Math NUM menu   Math FRAC menu

Scrolling a menu

After the number 9 in the first two pictures in Figure 1-6, a down arrow indicates that more items are available in the menu than appear on-screen. There's no down arrow after the 4 in the third screen in Figure 1-6 because that menu has exactly four items.

To see menu items that don't appear on-screen, repeatedly press ▼.

To get quickly to the bottom of a menu from the top of the menu, press ▲. Similarly, to quickly get from the bottom to the top, press ▼.

Selecting menu items

To select a menu item from a menu, key in the number (or letter) of the item or use the ▼▲ keys to highlight the number (or letter) of the item and then press ENTER.

Some menus, such as the Mode menu shown in the first screen in the upcoming Figure 1-7, require that you select an item from a list of items by highlighting that item. The list of items usually appears in a single row and the calculator requires that one item in each row be highlighted. To highlight an item, use the ▶◀▲▼ keys to place the cursor on the item and then press ENTER to highlight the item. The selections on the Mode menu are described in the next section.

To access Catalog Help, scroll to the menu item you want to use and press +. A screen showing the syntax of the command is displayed.

Setting the Mode

The Mode menu, which is accessed by pressing MODE, is the most important menu on the calculator; it tells the calculator how you want numbers and graphs to be displayed. The Mode menu for the TI-84 Plus C is pictured in the first screen in Figure 1-7.

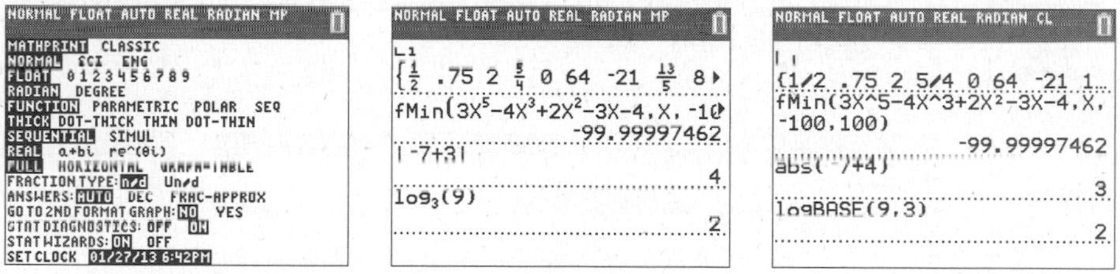

Figure 1-7: Mode, MathPrint, and Classic screens.

Mode menu   MathPrint mode   Classic mode

MathPrint mode versus Classic mode

The first choice on the Mode menu will have a big impact on the way your calculator displays expressions and answers. MathPrint is the default mode, and I strongly endorse using MathPrint at all times.

✔ **MathPrint mode:**

Fractions display like fractions, exponents look like exponents, text doesn't wrap to the next line, and templates make it easier to enter commands. See the second screen in Figure 1-7.

✔ **Classic mode:**

Fractions use a forward slash (/) symbol, most exponents aren't elevated, text wraps to the next line, and templates aren't available. See the third screen in Figure 1-7.

One item in each row of this menu must be selected. Here are your choices:

✔ **Normal, Sci, or Eng:**

This setting controls how numbers are displayed on the calculator. In Normal mode, the calculator displays a number in the usual numeric fashion that you used in elementary school — provided it can display it using no more than ten digits. If the number requires more than ten digits, the calculator displays it using scientific notation.

In Scientific (**Sci**) mode, numbers are displayed using scientific notation; and in Engineering (**Eng**) mode, numbers are displayed in engineering notation. These three modes are illustrated in Figure 1-8. In this figure, the first answer is displayed in normal notation, the second in scientific notation, and the third in engineering notation.

In scientific and engineering notation, the calculator uses **En** to denote multiplication by 10^n.

Figure 1-8:
Normal, scientific, and engineering notations.

| NORMAL FLOAT AUTO REAL RADIAN MP |
| 50*2000 |
| 100000 |

Normal mode

| SCI FLOAT AUTO REAL RADIAN MP |
| 50*2000 |
| 1ᴇ5 |

Scientific mode

| ENG FLOAT AUTO REAL RADIAN MP |
| 50*2000 |
| 100ᴇ3 |

Engineering mode

✔ **Float 0123456789:**

Select **Float** if you want the calculator to display as many digits as possible. Select **0** if you want all numbers rounded to an integer. If you're dealing with money, select **2** so that all numbers will be rounded to two decimal places. Selecting **5** rounds all numbers to five decimal places, and, well, you get the idea.

✔ **Radian or Degree:**

If you select **Radian**, all angles entered in the calculator are interpreted as being in radian measure; all angular answers given by the calculator will also be in radian measure. Similarly, if you select **Degree**, any angle you enter must be in degree measure, and any angular answer given by the calculator is also in degree measure.

✔ **Function, Parametric, Polar, or Seq:**

This setting tells the calculator what type of functions you plan to graph. Select **Function** to graph plain old vanilla functions in the form $y = f(x)$. Select **Parametric** to graph parametric equations; **Polar** to graph polar equations; and **Seq** to graph sequences. (Sequences are also called *iterative equations.*)

✔ **Thick, Dot–Thick, Thin, or Dot–Thin:**

In **Dot–Thick or Dot–Thin** mode, the calculator produces a graph by plotting only the points it calculates. In **Thick or Thin** mode, the calculator joins consecutively plotted points with a line. Thick or Thin has to do with the thickness of the line style in the Y= editor.

My recommendation is to select the **Thick or Thin** mode because each of the graphing options (**Function, Parametric, Polar,** and **Seq**) enables you to select a graphing style with the options of **Dot–Thick** or **Dot–Thin** line style.

If you want to quickly change the line styles of all of your functions at once, choose **Thick, Dot–Thick, Thin,** or **Dot–Thin.**

✔ **Sequential or Simul:**

In **Sequential** mode, the calculator completes the graph of one function before it graphs the next function. In Simultaneous (**Simul**) mode, the calculator graphs all functions at the same time. It does so by plotting the values of all functions for one value of the independent variable, and then plotting the values of all functions for the next value of the independent variable.

Simul mode is useful if you want to see whether two functions intersect at the same value of the independent variable. You have to watch the functions as they are graphed in order to *see* if this happens.

✔ **Real, a + b*i*, or re^θ*i*:**

If you're dealing with only real numbers, select the **Real** mode. If you're dealing with complex numbers, select **a + b*i*** if you want the complex numbers displayed in rectangular form. If you want complex numbers displayed in polar form, select the **re^θ*i*** mode.

✔ **Full, Horizontal, or Graph-Table:**

The **Full** screen mode displays the screen as you see it when you turn the calculator on. The other screen modes are split-screens. The **Horizontal** mode is for when you want to display a graph and the Y= editor or the Home screen at the same time. Use the **Graph-Table** mode when you want to display a graph and a table at the same time. (The split-screen modes are explained in detail in Chapters 9 and 10.)

✔ **Fraction Type: n/d or Un/d:**

The results display as simple fractions or mixed numbers.

✔ **Answers: Auto, Dec, Frac-Approx:**

Changing this setting affects how the answers are displayed. Choosing **Auto** displays answers in a similar form as the input. **Dec** displays answers in decimal form. **Frac-Approx** displays answers in fraction form when possible.

✔ **Go to 2nd Format Graph: NO, YES:**

Choosing Yes redirects you to the Format Graph screen. Alternatively, you can press [2nd][ZOOM].

✔ **Stat Diagnostics: OFF, ON:**

I recommend turning this ON so that r and r^2 display when you run a regression. See the first screen in Figure 1-9.

✔ **Stat Wizards: ON, OFF:**

If you have this set to ON, an input screen provides syntax help for entering the proper syntax of certain statistical commands. See the second screen in Figure 1-9.

✔ **Set Clock:**

This is where you set the clock on the TI-84 Plus family of calculators. To do this, use the arrow keys to place the cursor on the **SET CLOCK** option and press [ENTER]. You see the third screen in Figure 1-9. You use the [▶][◀][▲][▼] keys to move from item to item. To select items in the first, fifth, and eighth rows, place the cursor on the desired item and press [ENTER] to highlight that item. To enter numbers in the other options, edit the existing number or press [CLEAR] and use the keypad to enter a new number. When you're finished setting the clock, save your settings by placing the cursor on **SAVE** and pressing [ENTER].

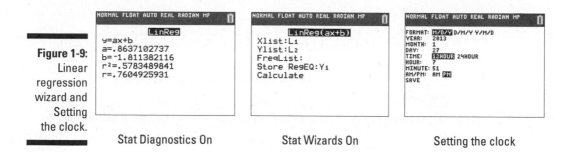

Figure 1-9:
Linear
regression
wizard and
Setting
the clock.

| Stat Diagnostics On | Stat Wizards On | Setting the clock |

REMEMBER

TIP

If you're planning on graphing trigonometric functions, put the calculator in Radian mode. Reason: Most trig functions are graphed for $-2\pi \le x \le 2\pi$. That's approximately $-6.28 \le x \le 6.28$. That's not a bad value for the limits on the x-axis. But if you graph in Degree mode, you would need $-360 \le x \le 360$ for the limits on the x-axis. This is doable . . . but trust me, it's easier to graph in Radian mode.

If your calculator is in Radian mode and you want to enter an angle in degrees, Chapter 3 tells you how to do so without resetting the mode.

REMEMBER

You can quickly check some of the mode settings (like radian or degree) by glancing at the status bar at the top of the screen.

Using the Catalog

The calculator's Catalog houses every command and function used by the calculator. However, it's usually easier to use the keyboard and the menus to access these commands and functions than it is to use the Catalog. There are several exceptions; for example, the hyperbolic functions are found only in the Catalog. If you have to use the Catalog, here's how to do it:

1. **If necessary, use the** ▷ ◁ △ ▽ **keys to place the cursor at the location where you want to insert a command or function found in the Catalog.**

   The command or function is usually inserted on the Home screen, or in the Y= editor when you're defining a function you plan to graph.

2. **Press** 2nd 0 **to enter the Catalog.**

   This is illustrated in the first screen in Figure 1-10.

3. **Enter the first letter in the name of the command or function.**

   Notice that the calculator is already in Alpha mode, as is indicated by the 🄰 in the upper-right part of the screen. To enter the letter, all you have to do is press the key corresponding to that letter. For example, if you're using the Catalog to access the function **seq(**, press LN because the letter **S** is written in green above this key. Use the ▾ key to scroll down to **seq(**. This is illustrated in the second screen in Figure 1-10.

4. **Repeatedly press ▾ to move the indicator to the desired command or function.**

5. **(Optional) Press ⊞ to access Catalog Help for the listed command or function.**

This is illustrated in the third screen in Figure 1-10. After pressing ENTER, the command or function is inserted at the cursor location.

6. **Press ENTER to select the command or function.**

After pressing ENTER, the command or function is inserted at the cursor location.

Figure 1-10:
Steps for
using the
Catalog.

NORMAL FLOAT AUTO REAL RADIAN MP	NORMAL FLOAT AUTO REAL RADIAN MP	NORMAL FLOAT AUTO REAL RADIAN MP
CATALOG	CATALOG	CATALOG HELP
▸abs(	2-SampZTest(	seq(■
and	Scatter	
angle(	Sci	(expression,variable
ANOVA(	Select(	,begin,end[,increment])
Ans	Send(	
Archive	▸seq(	
Asm(	Seq	
AsmComp(	Sequential	
Asm84CPrgm	setDate(	PASTE ESC

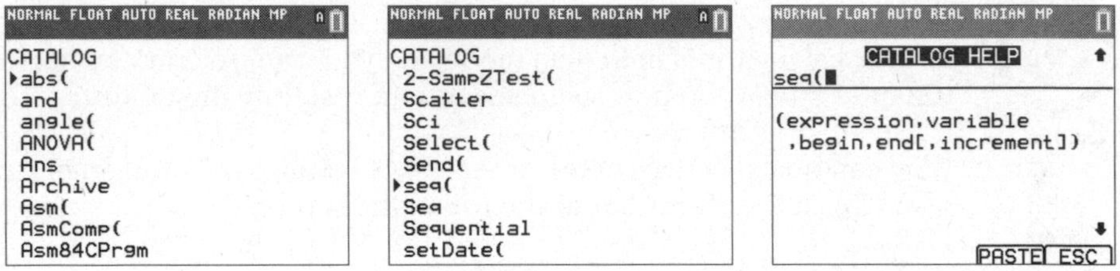

Press 2nd 0   Enter first letter   Catalog Help

Chapter 2

Doing Basic Arithmetic

· ·

In This Chapter

▶ Entering and evaluating arithmetic expressions

▶ Utilizing exponent and roots

▶ Working in scientific notation

▶ Knowing the important keys

▶ Obeying the order of operations

▶ Storing and recalling variables

▶ Using the previous answer

▶ Combining expressions

· ·

*W*hen you use the calculator to evaluate an arithmetic expression such as $5^{10} + 4^6$, the format in which the calculator displays the answer depends on how you have set the mode of the calculator. Do you want answers displayed in scientific notation? Do you want all numbers rounded to two decimal places?

Setting the *mode* of the calculator affords you the opportunity to tell the calculator how you want these — and other questions — answered. (Setting the mode is explained in Chapter 1.) As a general rule of thumb, highlight all the choices on the left side of the mode screen (refer to the first screen back in Figure 1-7).

Entering and Evaluating Expressions

Arithmetic expressions are evaluated on the Home screen. The Home screen is the screen you see when you turn the calculator on. If the Home screen is not already displayed on the calculator, press 2nd MODE to display it. If you want to clear the contents of the Home screen, repeatedly press CLEAR until the screen is empty.

Repeatedly pressing CLEAR doesn't delete your previous entries or answers — it just removes them from view! Press ▲ to scroll through your previous calculations. When a previous entry or answer is highlighted, press ENTER to paste it into your current entry line.

Arithmetic expressions are entered in the calculator the same way you would write them on paper. If you use the division sign (/) for fractional notation, it's usually a good idea to use parentheses around the numerator or the denominator, as illustrated in the first two calculations in Figure 2-1.

Figure 2-1:
Evaluating
arithmetic
expressions.

```
NORMAL FLOAT AUTO REAL RADIAN MP
5+3/2
                              6.5
(5+3)/2
                                4
```

There is a major difference between the subtraction key (─)and the negation key (⊡).They are not the same (see Figure 2-2), nor are they interchangeable. Use the ─ key to indicate subtraction; use the ⊡ key before a number to identify that number as negative.

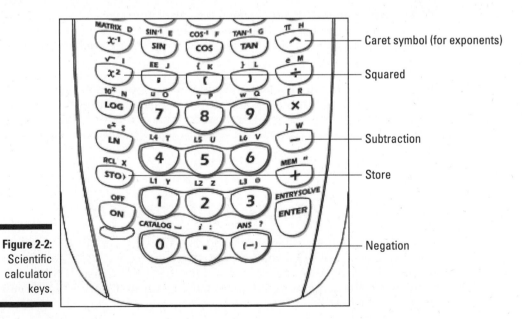

Figure 2-2:
Scientific
calculator
keys.

- Caret symbol (for exponents)
- Squared
- Subtraction
- Store
- Negation

If you improperly use ⊟ to indicate a subtraction problem, or if you improperly use ⊟ to indicate that a number is negative, you get the ERR: SYNTAX error message. The second screen in Figure 2-3 is the result of pressing ENTER on the first screen in Figure 2-3. Simply press ② to automatically bring your cursor to the cause of your error where you can edit the entry as needed.

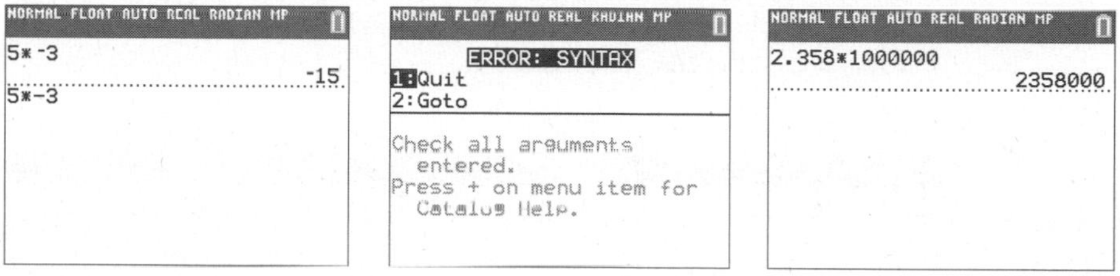

Figure 2-3: Entering calculations.

Subtraction and Negation   Error message   Entering large numbers

Do not use commas when entering numbers. For example, the number 1,000,000 is entered in the calculator as 1000000. See the third screen in Figure 2-3.

After entering the expression, press ENTER to evaluate it. The calculator displays the answer on the right side of the next line.

Using Exponents and Roots

In MathPrint mode, exponents actually look like exponents! There are two ways to square a number. One way is to type a number and press x^2. The advantage of using this technique is your cursor stays on the base of the Entry line (see the first screen in Figure 2-4). Another way of squaring a number is to type a number and press ^ ② to put the number to the second power. Notice that as soon as you press ^ (caret symbol), the cursor moves up to the exponent position. Type a number in the exponent position and the cursor will contain a small right arrow to remind you to press ▶ to bring the cursor back down to the base of the Entry line (see the second screen in Figure 2-4). Trust me when I tell you that it's really easy to forget to do this!

Of course, pressing ^ allows you to put a number to any exponent you would like (including negative exponents and rational exponents). See the first two lines of the last screen in Figure 2-4. There are two convenient secondary keys to perform commonly used exponential functions. Press 2nd LOG to produce the 10^x function, and press 2nd LN to generate the e^x function.

Square roots work in a similar fashion to exponents. Press 2nd x^2 to select a square root and type the expression you would like to evaluate. Notice that the cursor will stay under the radical sign until you press ▶ (see the last line of the third screen in Figure 2-4). To enter a root (other than a square root), press MATH 5 to access the $^x\sqrt{}$ template. Simply type the index and use the ▶ key to enter the expression you would like to evaluate. Alternatively, save time by typing the index first, access the $^x\sqrt{}$ template by pressing MATH 5, and then typing the expression. If you have a TI-84, be sure to type the index before accessing the $^x\sqrt{}$ template.

Figure 2-4:
Powers and
roots.

Cursor on the Base line   Cursor in the Exponent line   The $^x\sqrt{}$ template

Working in Scientific Notation

Scientific notation on a calculator looks a little different than what you're used to seeing in class. For example, $2.53*10^{12}$ will display as 2.53_E12. You can enter an expression in scientific notation by pressing 2nd , to type an $_E$, but entering an expression in scientific notation doesn't guarantee that your answer will remain in scientific notation. See the first screen in Figure 2-5.

In Normal mode, results that have a power of ten that are more than 9 or less than –3 are automatically expressed in scientific notation. In other words, any number that is more than ten digits or smaller than 0.001 will display in scientific notation. See the second screen in Figure 2-5.

Fortunately, you can force your calculator to display answers in scientific notation. Press MODE and use the arrow keys to choose Sci (short for Scientific mode) and press ENTER. You can rest assured that all your answers will be displayed in scientific notation. See the third screen in Figure 2-5.

It's not a good idea to leave your calculator in Sci mode. Doing so will not harm your calculator, but seeing every calculation in scientific notation may cause you to lose your mind!

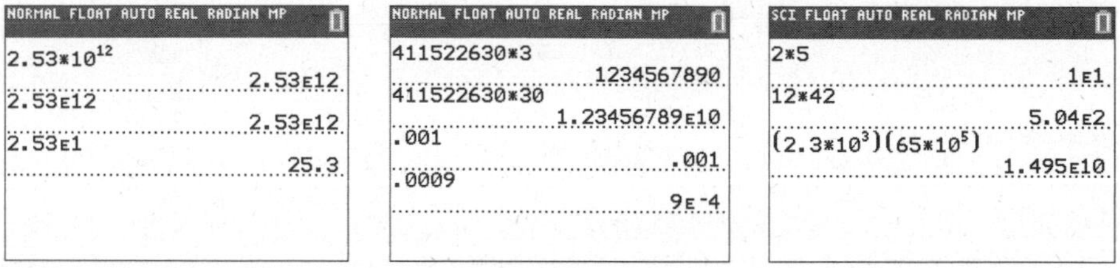

NORMAL FLOAT AUTO REAL RADIAN MP	NORMAL FLOAT AUTO REAL RADIAN MP	SCI FLOAT AUTO REAL RADIAN MP
2.53*10¹²	411522630*3	2*5
2.53ᴇ12	1234567890	1ᴇ1
2.53ᴇ12	411522630*30	12*42
2.53ᴇ12	1.23456789ᴇ10	5.04ᴇ2
2.53ᴇ1	.001	(2.3*10³)(65*10⁵)
25.3	.001	1.495ᴇ10
	.0009	
	9ᴇ-4	

Figure 2-5:
Scientific
notation.

      Accessing ᴇ   Using Normal mode   Using Sci mode

Getting Familiar with Important Keys

Starting with the fifth row of the calculator, you find the functions commonly used on a scientific calculator. Here's what they are and how you use them:

✔ **π and e**

The transcendental numbers π and e are respectively located in the fifth and sixth rows of the last column of the keyboard. To enter π in the calculator, press [2nd][^]; to enter e, press [2nd][÷], as shown in the first screen of Figure 2-6.

✔ **The trigonometric and inverse trigonometric functions**

The trigonometric and inverse trigonometric functions are located in the fifth row of the keyboard. These functions require that the argument of the function be enclosed in parentheses. To remind you of this, the calculator provides the first parenthesis for you (as shown in the second screen of Figure 2-6).

✔ **The inverse function**

The inverse function is located in the fifth row of the left column on the calculator. To enter the multiplicative inverse of a number, enter the number and press [x⁻¹]. When dealing with matrices, using the [x⁻¹] key will calculate the inverse of the matrix (see Chapter 8 for more about matrices). The third screen in Figure 2-6 shows these operations.

If you want to evaluate an arithmetic expression and you need a function other than those just listed, you'll most likely find that function in the Math menu (described in detail in Chapter 6).

You can impress your friends at parties by pointing out that TI-84 Plus uses 3.1415926535898 for π in calculations.

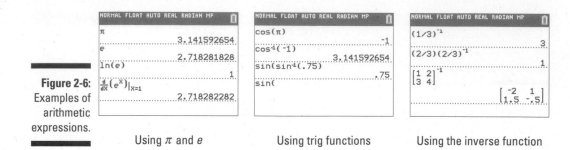

Figure 2-6:
Examples of
arithmetic
expressions.

Using π and e Using trig functions   Using the inverse function

Following the Order of Operations

The order in which the calculator performs operations is the standard order
that you are used to. Spelled out in detail, here is the order in which the calculator performs operations:

1. **The calculator simplifies all expressions surrounded by parentheses.**

2. **The calculator evaluates all functions that are followed by the**
   argument.

   These functions supply the first parenthesis in the pair of parentheses
   that must surround the argument. An example is sin x. When you press
   SIN to access this function, the calculator inserts **sin(** on-screen. You
   then enter the argument and press).

3. **The calculator evaluates all functions entered after the argument.**

   An example of such a function is the square function. You enter the
   argument and press x^2 to square it.

   Evaluating -3^2 may not give you the expected answer. You think of -3
   as being a single, negative number. So when you square it, you expect
   to get $+9$. But the calculator gets -9 (as indicated in the first screen of
   Figure 2-7). This happens because the normal way to enter -3 into the
   calculator is by pressing (-)3 — and pressing the (-) key is equivalent
   to multiplying by -1. Thus, in this context, $-3^2 = -1 * 3^2 = -1 * 9 = -9$. To
   avoid this potentially hazardous problem, always surround negative
   numbers with parentheses *before* raising them to a power. See the first
   screen in Figure 2-7.

4. **The calculator evaluates powers entered using the** ⌃ **key and roots**
   entered using the $^x\sqrt{}$ **function.**

   The $^x\sqrt{}$ function is found in the Math menu. You can also enter various
   roots by using fractional exponents — for example, the cube root of 8 can
   be entered by pressing 8⌃1÷3. See the second screen in Figure 2-7.

5. **The calculator evaluates all multiplication and division problems as it encounters them, proceeding from left to right.**

6. **The calculator evaluates all addition and subtraction problems as it encounters them, proceeding from left to right.**

NORMAL FLOAT AUTO REAL RADIAN MP

$(-3)^2$

       9

-3^2

       -9

Using parentheses

NORMAL FLOAT AUTO REAL RADIAN MP

$-8^{1/3}$

       -2

$8^{1/3}$

       2

$9^{1/2}$

       3

Exponent roots

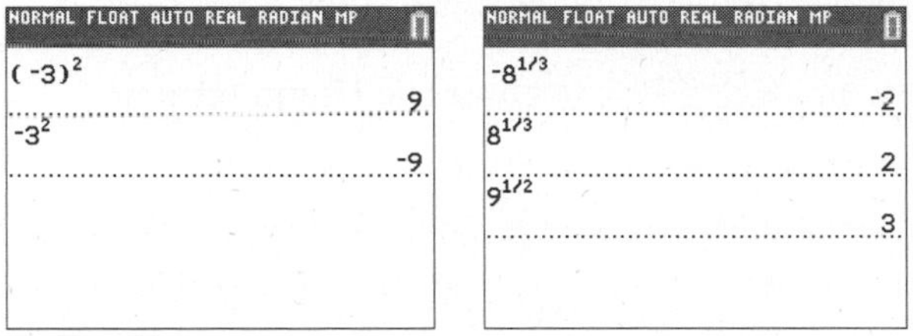

Figure 2-7: Order of operations.

Using the Previous Answer

You can use the previous answer in the next arithmetic expression you want to evaluate. If that answer is to appear at the beginning of the arithmetic expression, first key in the operation that is to appear after the answer. The calculator displays **Ans** followed by the operation. Then, key in the rest of the arithmetic expression and press ENTER to evaluate it. See the first screen in Figure 2-8. Pressing ENTER repeatedly will recycle the last entry and generate a sequence of numbers. See the second screen in Figure 2-8.

If you want to embed the last answer in the next arithmetic expression, key in the beginning of the expression to the point where you want to insert the previous answer. Then press 2nd (-) to key in the last answer. Finally, key in the rest of the expression and press ENTER to evaluate it. Pressing ENTER repeatedly will generate a sequence. See the third screen in Figure 2-8.

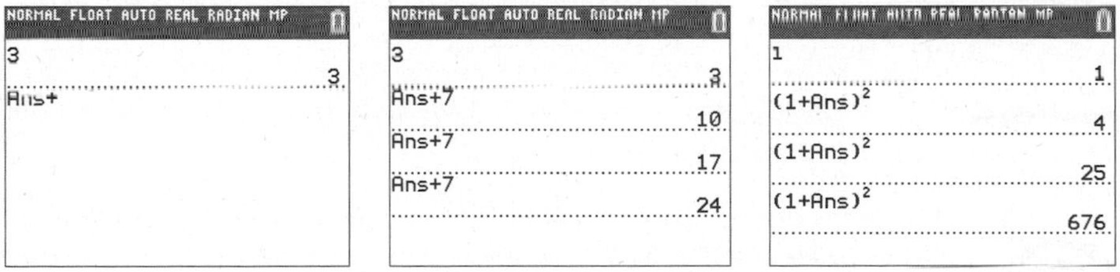

NORMAL FLOAT AUTO REAL RADIAN MP

3

       3

Ans+

Press +

NORMAL FLOAT AUTO REAL RADIAN MP

3

       3

Ans+7

       10

Ans+7

       17

Ans+7

       24

Pressing ENTER

NORMAL FLOAT AUTO REAL RADIAN MP

1

       1

$(1+Ans)^2$

       4

$(1+Ans)^2$

       25

$(1+Ans)^2$

       676

Embedding the last answer

Figure 2-8: Using the previous answer.

Storing Variables

The letters STO may look like texting language, but the [STO▸] key on a calculator is a handy feature to have around. If you plan to use the same number many times when evaluating arithmetic expressions, consider storing that number in a variable. To do so, follow these steps:

1. **If necessary, press [2nd][MODE] to enter the Home screen.**

2. **Enter the number you want to store in a variable.**

   You can store a number or an arithmetic expression.

3. **Press [STO▸].**

   The result of this action is shown in the first screen in Figure 2-9.

4. **Press [ALPHA] and press the key corresponding to the letter of the variable in which you want to store the number.**

   The letters used for storing variables are the letters of the alphabet and the Greek letter θ.

5. **Press [ENTER] to store the value.**

   This is illustrated in the second screen in Figure 2-9.

After you have stored a number in a variable, you can insert that number into an expression. To do so, place the cursor where you want the number to appear, press [ALPHA], and press the key corresponding to the letter of the variable in which the number is stored. See the third screen in Figure 2-9.

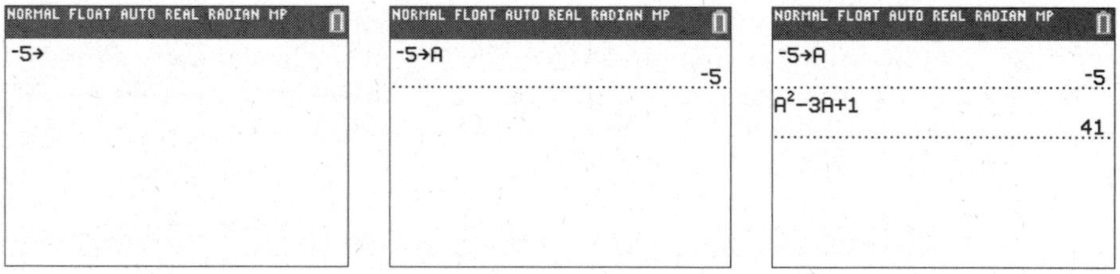

NORMAL FLOAT AUTO REAL RADIAN MP	NORMAL FLOAT AUTO REAL RADIAN MP	NORMAL FLOAT AUTO REAL RADIAN MP
-5▸	-5→A <div align=right>-5</div>	-5→A <div align=right>-5</div> A²-3A+1 <div align=right>41</div>

Figure 2-9:
Storing
steps.

Press [STO▸]   Enter variable   Using a stored variable

The number you store in a variable remains stored in that variable until you *or the calculator* stores a new number in that variable. Because the calculator uses the letters X, T, and θ when graphing functions, parametric equations, and polar equations, it is possible that the calculator will change the value stored in these variables when the calculator is in graphing mode. For example, if you store a number in the variable X and ask the calculator to find the zero of the graphed function X^2, the calculator will replace the number stored in X with 0, the zero of X^2. So avoid storing values in these three variables if you want that value to remain stored in that variable after you have graphed functions, parametric equations, or polar equations.

Combining Expressions

You can *combine* (link) several expressions or commands into one expression by using a colon to separate the expressions or commands. The colon is entered into the calculator by pressing ALPHA . . See the first screen in Figure 2-10. Combining expressions is not a timesaver, but it is a space saver. For comparison, see the second screen in Figure 2-10, where the expressions were not combined into one line.

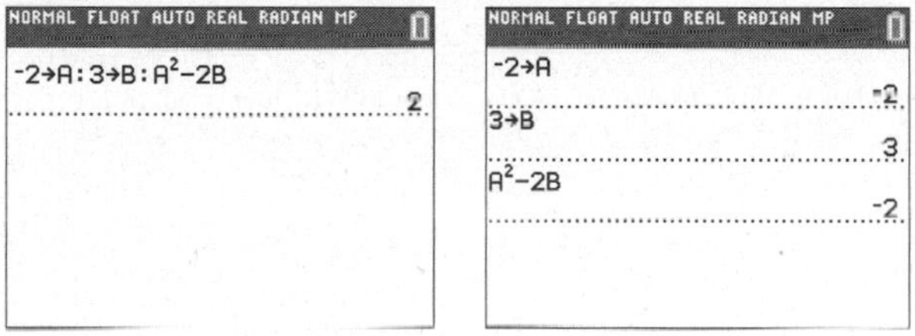

Figure 2-10: Combining expressions.

Combining expressions   Separate expressions

Chapter 3

Dealing with Fractions

* * *

In This Chapter

▶ Setting the mode

▶ Converting fractions and decimals

▶ Accessing hidden shortcut menus

▶ Using fractions and mixed numbers

▶ Entering complex numbers in fractions

▶ Working with complex fractions

* * *

1 often hear students ask, "Where is the fraction key?" The short answer is that there's no fraction key, per se. The long answer is that there are many fraction tools built into this calculator. For starters, isn't a fraction just division in disguise? So, pressing ÷ between two numbers creates a fraction. Of course, there's much more to dealing with fractions on this calculator. To learn all the fraction functionality that is at your fingertips, just continue reading this chapter.

Setting the Mode

Do you prefer fractions or decimals? Would you rather work with an improper fraction or a mixed number? There's no right answer to these questions, but what would make you (or your teacher) happy? Changing the mode of your calculator forces the calculated answers into a form of your liking. Be careful; this is a big decision on your part! Setting the mode not only affects calculations on the Home screen, but also the way lists and sequences are displayed.

To change the form of your calculated answers, press MODE. Use the arrow keys to scroll to the 11th line, ANSWERS. Here, there are three choices that affect how calculated answers are displayed:

✔ **AUTO:** Choosing AUTO displays the answers in a similar format to the way the expressions are entered. If the expression contains a decimal, then you should expect the answer to be in decimal form. If the expression is entered in fraction form, then you should expect the answer to be expressed in fraction form. See the first screen in Figure 3-1.

✔ **DEC:** The DEC mode forces the answers to be displayed as decimals. See the second screen in Figure 3-1.

✔ **FRAC — APPROX:** When possible, the FRAC — APPROX mode displays answers as fractions. See the third screen in Figure 3-1.

NORMAL FLOAT AUTO REAL RADIAN MP	NORMAL FLOAT DEC REAL RADIAN MP	NORMAL FLOAT FRAC REAL RADIAN MP
1/2+1/4 .75	1/2+1/4 .75	1/2+1/4 $\frac{3}{4}$
$\frac{1}{2}+\frac{1}{4}$ $\frac{3}{4}$	$\frac{1}{2}+\frac{1}{4}$ .75	$\frac{1}{2}+\frac{1}{4}$ $\frac{3}{4}$
$\frac{1.0}{2}+\frac{1}{4}$ .75	$\frac{1.0}{2}+\frac{1}{4}$ .75	$\frac{1.0}{2}+\frac{1}{4}$ $\frac{3}{4}$
AUTO mode	DEC mode	FRAC–APPROX mode

Figure 3-1: Modes of the calculated answer.

There's one more mode decision you need to make. What type of fraction do you prefer: improper fractions or mixed numbers? Press MODE and change the FRACTION TYPE to one of these two choices:

✔ **n/d:** Fractions are displayed in simplified fraction form.

The numerator of a fraction must contain less than seven digits and the denominator of a fraction must not exceed 9999.

✔ **Un/d:** When possible, fractions are displayed as a mixed number.

To avoid errors and potential problems, enter *U*, *n*, and *d* as integers with a maximum of three digits.

Converting Fractions and Decimals

There's an easy way to convert a decimal to a fraction, regardless of the mode setting. You can access the **Frac** and **Dec** functions in the first two options in the Math menu. The **Frac** function displays an answer as a fraction. Type the expression and press MATH ENTER ENTER to display the expression as a fraction. Often, I don't think ahead and my answer is a decimal (when I wanted a fraction.) No problem! Press [MATH][ENTER][ENTER] and your answer is converted to a fraction. See the first screen in Figure 3-2. If your calculator can't convert an expression to a fraction, it lets you know by redisplaying the decimal.

How do you convert an infinite repeating decimal into a fraction? Just type at least ten digits of the repeating decimal and press [MATH][ENTER][ENTER]. See the second screen in Figure 3-2.

The **Dec** function converts a fraction to a decimal. Enter the fraction and press [MATH][2][ENTER]. Of course, if you're not thinking ahead and your answer is in fraction form, just press [MATH][2][ENTER] to display your answer as a decimal. An example is shown in the third screen in Figure 3-2.

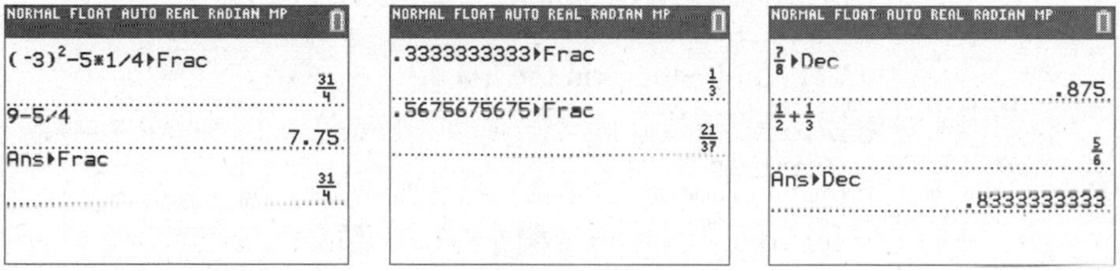

Figure 3-2:
Converting
fractions
and
decimals.

| Frac function | Repeating decimals | Dec function |

Accessing Shortcut Menus

Did you know that there are four hidden shortcut menus on your calculator? The four menus are: FRAC (Fraction menu), FUNC (Function menu), MTRX (Matrix menu), and YVAR (Y-variables menu). To access the hidden FRAC menu, press [ALPHA][Y=]. See the first screen in Figure 3-3. Notice that after pressing [ALPHA], the keys at the top of your keypad become soft keys that activate on-screen menus.

The MTRX menu can <u>only</u> be accessed by pressing [ALPHA][ZOOM] to access the MTRX shortcut menu. However, the rest of the shortcut menus can also be accessed by standard menus. For example, the FRAC menu can also be accessed in two places in the MATH menu. Press [MATH][◄] or press [MATH][►][▲] (at the bottom of the NUM menu) to find the FRAC menu in a standard menu. See the second and third screens in Figure 3-3.

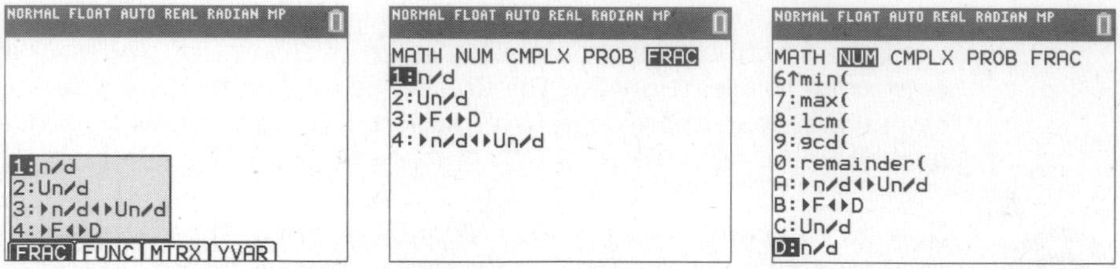

Figure 3-3: Accessing the FRAC menu.

Hidden menus   FRAC menu in MATH menu   FRAC menu in NUM menu

Entering Fractions and Mixed Numbers

Press [ALPHA][Y=] to access the FRAC menu. The first two options in the FRAC menu are easy-to-use fraction templates:

- **n/d:** Enter fractions in the fraction template.

- **Un/d:** Enter fractions in the mixed number template. See the first screen in Figure 3-4.

The next two options are used for conversion:

- **▸n/d◂▸Un/d:** Converts a mixed number to an improper fraction, or an improper fraction to a mixed number. See the second screen in Figure 3-4.

- **▸F◂▸D:** Converts a fraction to a decimal, or vice versa. See the third screen in Figure 3-4.

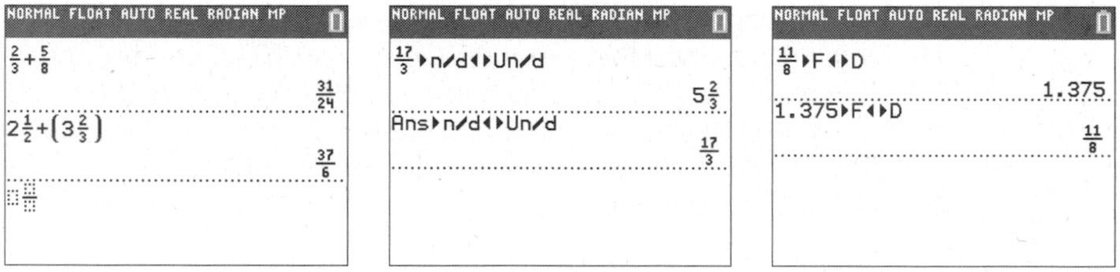

Figure 3-4: Fraction templates and conversions tools.

n/d and Un/d templates   Converting mixed numbers   Fractions and decimals

Entering Complex Numbers in Fractions

As much as I love the n/d fraction template, it has its limitations. Complex numbers may not be used in the n/d fraction template. To enter the complex number, *i*, press [ALPHA][.]. (For more about working with complex numbers, see Chapter 5.) Entering a complex number in the n/d fraction template produces an error message as seen in the first screen in Figure 3-5.

Don't worry! You can enter complex numbers into fractions the old-fashioned way, using parentheses and the [÷] key. Your calculator automatically simplifies fractions that contain a complex number in the denominator. See the second screen in Figure 3-5.

Figure 3-5:
Entering complex numbers in fractions.

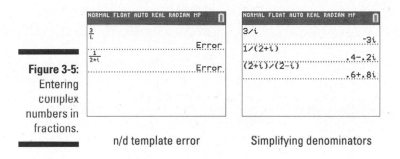

n/d template error   Simplifying denominators

Entering Complex Fractions

Complex fractions are fractions that contain one or more fractions in the numerator or denominator. In other words, complex fractions have fractions inside of fractions. No problem. Complex fractions can easily be entered in your calculator by using the n/d fraction template multiple times in the same fraction. See Figure 3-6.

Press [ALPHA][Y=] to access the n/d fraction template.

Figure 3-6:
Entering complex fractions.

Chapter 4

Solving Equations

. . . .

In This Chapter

▶ Entering, editing, and solving equations in the Equation Solver

▶ Guessing the value to find multiple solutions

▶ Using the Solve function

▶ Finding roots of polynomials

▶ Solving systems of equations

. . . .

Many students don't know what a powerful tool their calculator is! You can use your calculator to solve all kinds of different equations. Three methods are discussed in this chapter: Equation Solver, Solve function, and the PlySmlt2 app. A fourth method of solving equations — graphing — is covered in Chapter 11.

One note that applies to all the methods of solving equations discussed in this chapter: Your calculator automatically displays irrational answers as decimals. For instance, the $\sqrt{2}$ will be displayed as 1.414213562. This can be problematic if you (or your teacher) want exact answers displayed. However, if you're taking a standardized test, you can easily check your answer by using your calculator to convert radical answers to a decimal.

Using the Equation Solver

The Equation Solver is a great tool for solving one-variable equations. The Solver is also capable of solving an equation for one variable given the values of the other variables. Keep in mind that the Solver can only produce real-number solutions.

The following lists the basic steps for using the Equation Solver. Each of these steps is explained in detail following this list. If you have never used the Equation Solver before, I suggest that you read the detailed explanations for each step because the Equation Solver is a bit tricky. After you have had experience using the Solver, you can refer back to this list, if necessary, to refresh your memory on its use.

1. **Enter a new equation in the Equation Solver.**

2. **Enter a guess for the solution.**

3. **Press** ALPHA ENTER **to solve the equation.**

Step 1: Enter or edit the equation to be solved

For this exercise, I'm going to use the Equation Solver to solve the equation, $2(3 - X) = 4X - 7$. To enter an equation in the Solver, follow these steps:

1. **Press** MATH ▲ ENTER **to access the Solver from the Math menu.**

   When the Solver appears, it should look similar to the first screen in Figure 4-1.

   The Solver in the TI-84 Plus works a little differently. Set the original equation equal to zero and enter the resulting equation into the Solver.

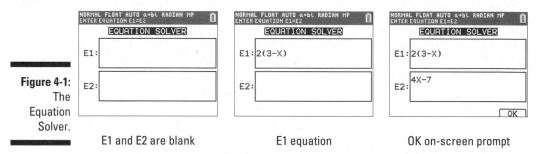

Figure 4-1:
The
Equation
Solver.

E1 and E2 are blank   E1 equation   OK on-screen prompt

2. **Enter the left side of the equation to be solved in E1.**

   If equation **E1** already contains an equation, press CLEAR before entering the left side of the equation to be solved. See the second screen in Figure 4-1.

3. **Press** ▼ **and enter the right side of the equation to be solved in E2.**

   If equation **E2** already contains an equation, press CLEAR before entering the right side of the equation to be solved.

4. **Press** GRAPH **to activate the on-screen OK prompt.**

Notice that the on-screen OK prompt does not appear until you enter expressions in both **E1** and **E2**. See the third screen in Figure 4-1.

You can also use a function that you've entered in the Y= editor in the definition of your equation. To insert such a function into the **E1** or **E2**, press ALPHA TRACE to access the Y-variables menu and then press the number of the Y-variable you want to enter. The Y= editor is explained in Chapter 9.

Step 2: Guess a solution

Guess at a solution. Any value in the interval defined by the **bound** variable will do. Guessing is necessary because your calculator solves problems through an iterative process. The **bound** variable at the bottom of the screen (see the first screen in Figure 4-2) is where you enter the bounds of the interval containing the solution you're seeking. The default setting for this interval is $[-10^{99}, 10^{99}]$, as is indicated by **bound** = {−1E99, 1E99}.

1E99 is $(1*10^{99})$ in scientific notation. That's a huge number!

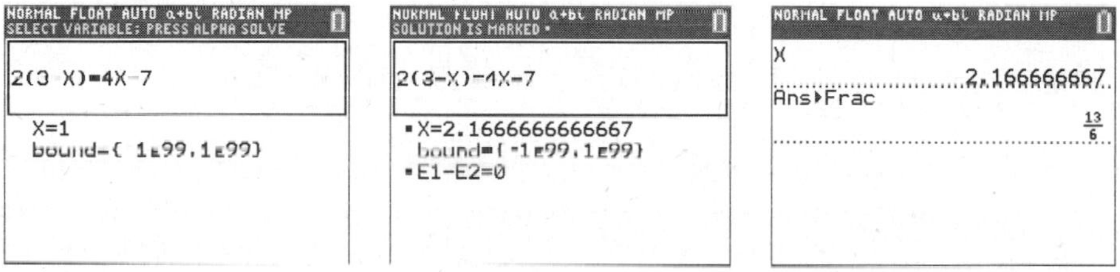

Figure 4-2: Steps for solving an equation in the Equation Solver.

NORMAL FLOAT AUTO a+bi RADIAN MP
SELECT VARIABLE: PRESS ALPHA SOLVE
 	
2(3-X)=4X-7
 	
X=1
bound={-1ε99,1ε99}
	NORMAL FLOAT AUTO a+bi RADIAN MP
	SOLUTION IS MARKED •
	
	2(3-X)-4X-7
	
	• X=2.1666666666667
	  bound={-1ε99,1ε99}
	• E1-E2=0
		NORMAL FLOAT AUTO a+bi RADIAN MP
		
		X
		          2.166666667
		Ans▶Frac
		             $\frac{13}{6}$

Bound variable	Press ALPHA ENTER	Answer as a fraction

If your guess is close to the solution, the calculator quickly solves the equation; if it's not, it may take the calculator a while to solve the equation. I usually guess 1 for an equation that has one solution.

If your equation has more than one solution, the calculator will find the one closest to your guess.

Step 3: Solve the equation

To solve an equation, follow these steps:

1. **Use the ▲▼ keys to place the cursor anywhere in the line that contains the variable you're solving for.**

   Place your cursor in the variable for which you want to make a guess.

2. **Press [ALPHA][ENTER] to solve the equation.**

   The second screen of Figure 4-2 shows this procedure; the square indicator shown next to the **X** indicates that **X** is the variable just solved for.

You can access the calculated solution on the Home screen. Press [2nd][MODE] to quit the application. Next, type the variable you solved for, in this case, X. Press [X,T,Θ,n][ENTER] to see your answer in decimal form. Press [MATH][ENTER][ENTER] to convert your answer to a fraction. See the third screen in Figure 4-2.

The **E1 – E2** value that appears at the bottom of the second screen in Figure 4-2 evaluates the two sides of the equation (using the values assigned to the variables) and displays the difference — that is, the accuracy of this solution. An **E1 – E2** value of zero indicates an exact solution.

If you get the ERR: NO SIGN CHNG error message when you attempt to solve an equation using the Equation Solver, then the equation has no real solutions in the interval defined by the **bound** variable.

Assigning Values to Variables

Did you know that your calculator can solve equations that have more than one variable? The trick is that you must assign values to all the variables except the one that you're solving for. For example, here is a classic math question:

> Find the equation of the line (in slope-intercept form) that goes through the point (–3, –4) and has a slope of 1/4.

To solve this problem, enter the equation Y = M*X + B into the Solver. Press [MATH][▲][ENTER] to access the Solver. Enter the equation into **E1** and **E2**. See the first screen in Figure 4-3.

After you have entered an equation in the Solver, the values assigned to the variables in your equation are the values that are currently stored in those variables in your calculator. You must assign an accurate value to all variables except the variable you're solving for. These values must be real numbers or arithmetic expressions that simplify to real numbers.

To assign a value to a variable, use the [▶][◀][▲][▼] keys to place the cursor on the number currently assigned to that variable and then key in the new value.

As you start to key in the new value, the old value is erased. Press ENTER when you're finished entering the new value (as illustrated in the second screen of Figure 4-3, where values are assigned to variables **Y**, **M**, and **X**).

Enter a guess for **B** and press ALPHA ENTER to solve the equation. Since you have found the value of **B**, you can answer the question and write the equation of the line, Y = 1/4X – 13/4.

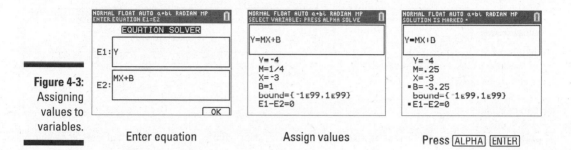

Figure 4-3:
Assigning
values to
variables.

Enter equation   Assign values   Press ALPHA ENTER

Finding Multiple Solutions

Some equations can have more than one solution. If you're dealing with an absolute value equation or an equation with a degree larger than one, there's a good chance the equation will have multiple solutions.

Using the Equation Solver, you can employ one of two techniques to find multiple solutions to equations.

Follow the three steps laid out in the first section of this chapter to use the Solver to solve an equation.

Making strategic guesses

When I expect multiple solutions, I usually guess a large positive number as my first guess. This strategy typically produces the largest solution for the equation you're solving. The first screen in Figure 4-4 is the result of guessing 100.

After I find one solution, I guess the opposite of the large positive number as my second guess. Many times, this technique will find the smallest solution. I guessed –100 for my second guess, as shown in the second screen in Figure 4-4.

The result of guessing –100 is pictured in the third screen in Figure 4-4. Depending on the equation, it may be necessary to continue guessing until you find all the solutions to the equation.

An **E1-E2** value of zero indicates an exact solution. The third screen in Figure 4-4 shows a solution that is off by the extremely small number ($-1.1*10^{-11}$).

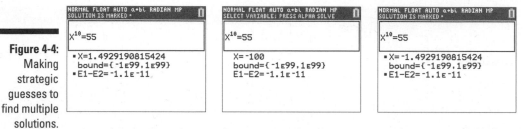

Figure 4-4: Making strategic guesses to find multiple solutions.

Result of guessing 100   Guessing –100   Result of guessing –100

Defining the solution bounds

When the equation you're solving has multiple solutions, it's sometimes help-ful to redefine the **bound** variable. Trigonometric functions are notorious for having infinite number of solutions. Often, a question will ask for the solu-tions within a certain interval. Adjusting the **bound** variable at the bottom of the screen assures that you will only find solutions that are in the interval defined by the **bound** variable.

Follow these steps to redefine the bound variable:

1. **Use the** ▶◀▲▼ **keys to place the cursor anywhere in the line contain-ing the bound variable.**

2. **Press** CLEAR **to erase the current entry.**

3. **Press** 2nd(**to insert the left brace.**

4. **Enter the lower bound, press** , **, enter the upper bound, and then press** 2nd) **to insert the right brace.**

5. **Press** ENTER **to store the new setting in the bound variable, or press** ▲ **to make your guess.**

Here is a typical question you might see:

Find all the real solutions to the function Y = 3*sin(2X + 1), where $0 < X < \pi$.

Enter the equation into **E1**. Change the bounds following the preceding steps. See the first screen in Figure 4-5.

Guess a number close to the lower bound. I guessed 0.3 and you can see the resulting solution in the second screen in Figure 4-5.

Now, guess a number close to the upper bound. I guessed 3 and was able to find another solution, as shown in the third screen in Figure 4-5.

If the variable you're solving for is assigned a value (guess) that isn't in the interval defined by the **bound** variable, then you get the ERR: BAD GUESS error message.

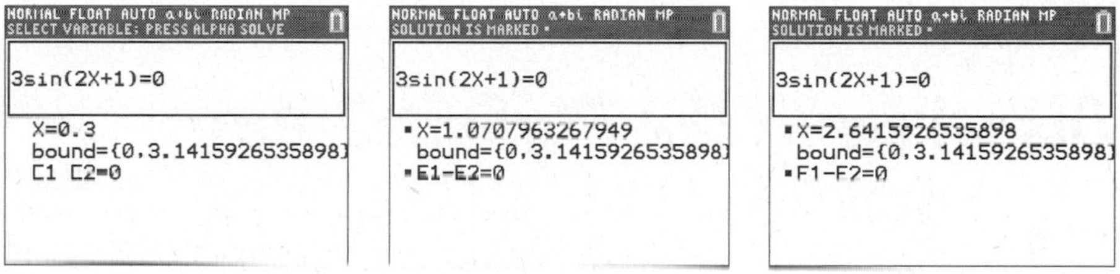

Figure 4-5:
Defining the
bound
variable.

NORMAL FLOAT AUTO a+bi RADIAN MP	NORMAL FLOAT AUTO a+bi RADIAN MP	NORMAL FLOAT AUTO a+bi RADIAN MP
SELECT VARIABLE: PRESS ALPHA SOLVE	SOLUTION IS MARKED •	SOLUTION IS MARKED •
3sin(2X+1)=0	3sin(2X+1)=0	3sin(2X+1)=0
X=0.3	•X=1.0707963267949	•X=2.6415926535898
bound={0,3.1415926535898}	bound={0,3.1415926535898}	bound={0,3.1415926535898}
E1 E2=0	•E1-E2=0	•F1-F2=0

Changing the bounds   Guessing 0.3   Guessing 3

Using the Solve Function

There are multiple ways to use the calculator to solve equations. The Solve function is difficult to locate, but relatively painless to use. Unfortunately, the Solve function can only be found in the catalog. Press [2nd][0] to access the catalog.

Alpha-lock is automatically on while viewing the catalog (as indicated by the flashing **A** in the cursor.) Pressing one of the many keys that have a letter jumps your cursor to the first item in the catalog that begins with the letter you pressed. For example, press [LN] to jump the items in the catalog that begin with the letter S.

Scroll to the Solve function and press [ENTER]. First, set the equation to be solved equal to zero. To solve X/2 + 5 = –2X, add 2X to both sides. The syntax of the Solve function is: Solve(expression, variable, guess). The expression is the part of an equation that has been set equal to zero. See the first screen in Figure 4-6.

A strategic guess allows you to solve equations that have more than one solution. I usually guess a large negative number on the first calculation followed by a large positive number as shown in the second screen in Figure 4-6.

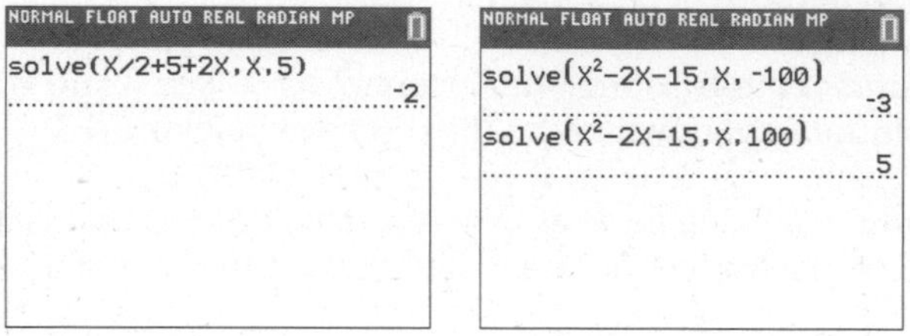

NORMAL FLOAT AUTO REAL RADIAN MP

solve(X/2+5+2X,X,5)
                                    -2

Solving linear equations

NORMAL FLOAT AUTO REAL RADIAN MP

solve(x²-2X-15,X,-100)
                                    -3
solve(x²-2X-15,X,100)
                                    5

Solving quadratic equations

Figure 4-6:
Using the
Solve
function.

Discovering the PlySmlt2 App

Using the Equation Solver or the Solve function works pretty well for linear or quadratic equations. But how can you use your calculator to solve polynomial equations with a degree bigger than two? Enter the PlySmlt2 app. Funny name. This app is truly multi-dimensional! Ply is short for Polynomial Root Finder. Smlt2 is short for Simultaneous Equation Solver. Unlike the Equation Solver and the Solve function, this app can find imaginary or complex solutions. Keep reading to find out how this powerful app makes solving equations a little easier.

Press APPS to access the list of apps that are pre-loaded on your calculator. Use ▲ to scroll to PlySmlt2 and press ENTER.

Finding the roots of a polynomial

Once the PlySmlt2 app has started, press 1 on the MAIN MENU to begin finding the roots of polynomials. First, configure the poly root finder mode screen. To find roots of the polynomial, $Y = X^3 + 3X^2 - 6X - 8$, I set the order (degree) to three. See the first screen in Figure 4-7.

On the MAIN MENU, press 4 to access the Help menu. Sometimes Help menus aren't that helpful, but here's what I learned from this Help menu: Real mode is not available in this app for polynomials with degree bigger than three. Press 2nd MODE to quit the Help menu.

Navigate to the next screen by using the five keys located on the top of your keypad. These keys are soft keys that select on-screen prompts. To advance to the next screen in the app, press GRAPH, right below the on-screen NEXT prompt.

Enter the coefficients on the next page. See the second screen in Figure 4-7. If the polynomial is missing a term, be sure to enter a zero for the missing coefficient(s). For instance, for the polynomial $Y = X^3 + 8$, enter the coefficients 1,0,0,8.

Finding the roots is easy: Just press GRAPH right below the on-screen SOLVE prompt. The roots of the polynomial are displayed in a vertical column. See the third screen in Figure 4-7.

I like two of the on-screen options on the page where the roots are displayed. Press TRACE to store the polynomial equation to Y=, or press GRAPH to convert fractions to decimals, and vice versa.

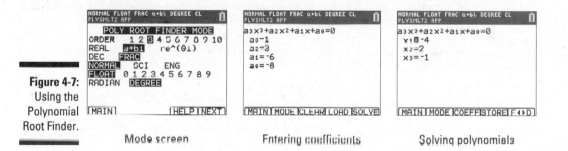

Figure 4-7: Using the Polynomial Root Finder.

Mode screen   Entering coefficients   Solving polynomials

Solving systems of equations

After starting the PlySmlt2 app, press 2 on the MAIN MENU screen to begin solving a system of equations. Configure the mode screen to match the system you're trying to solve. See the first screen in Figure 4-8.

Navigate to the next screen by using the five keys located on the top of your keypad. These keys are soft keys that select on-screen prompts. To advance to the next screen in the app, press GRAPH, right below the on-screen NEXT prompt.

I chose three equations with three unknowns in order to solve the following system:

$$2A + 3B - 2C = 8$$
$$A \quad - 4C = 1$$
$$2A - \ B - 6C = 4$$

It's important to organize your system so that your coefficients are aligned vertically and your constants are on the right side of each equation in your system. Your coefficients and constants must be entered in augmented matrix form. You may notice a line in the matrix that separates the coefficients from the constants (where the equal sign in each equation is located.) See the second screen in Figure 4-8 to see how I entered my system of equations.

Solving the system of equations is easy: Just press GRAPH right below the on-screen SOLVE prompt. The solutions of the system are displayed in a vertical column. See the third screen in Figure 4-8. Press 2nd MODE to quit the PlySmlt2 app.

Figure 4-8:
Using the
Simulta-
neous
Equation
Solver.

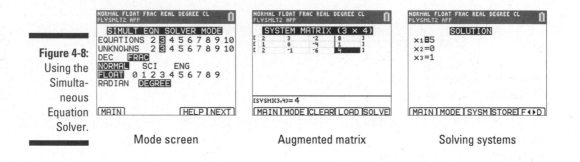

Mode screen   Augmented matrix   Solving systems

Part II
Taking Your Calculator Relationship to the Next Level

```
NORMAL FLOAT AUTO re^(θi) RADIAN MP

[1 2 3] + [-1 2 -4]
[4 5 6]   [ 8 3 -1]
                        [ 0 4 -1]
                        [12 8  5]
..........................................
[1 2 3] - [-1 2 -4]
[4 5 6]   [ 8 3 -1]
                        [ 2 0 7]
                        [-4 2 7]
..........................................
```

web extras Have a look-see at ways you can use your calculator to solve standardized test questions at www.dummies.com/extras/ti84plus.

In this part . . .

✔ See how to enter and work with complex numbers.

✔ Explore the dozens of commands found in the Math menu.

✔ Learn to use the commands in the Angle menu to evaluate expressions in both Radian and Degree mode.

✔ Familiarize yourself with how to use the Boolean logic feature to your advantage.

✔ Find out how to enter matrices and use them to solve a system of equations.

Chapter 5

Working with Complex Numbers

*E*arly on in your math journey, you were probably told that you can't take the square root of a negative number. Then a teacher blew your mind by saying you really can take the square root of a negative number and the result will contain the imaginary number, *i*. Complex numbers are of the form *a* + *bi*, where *a* is the real part and *b* is the imaginary part. Fortunately, your calculator knows how to handle complex numbers. In fact, there's a CMPLX menu of functions on your calculator designed to accomplish just about any task you need to when working with complex numbers.

Setting the Mode

Try evaluating $\sqrt{-1}$ in your calculator. On the Home screen, press [2nd][x^2][(-)][1] [ENTER]. There's a good chance you'll get an ERROR: NONREAL ANSWERS message, as shown in the first screen in Figure 5-1.

In Real mode, your calculator usually returns an error for a complex-number result. The exception is when you enter your expression using *i*. In this case, your calculator produces a complex-number result regardless of the mode. The good news is you can avoid this error altogether by setting the mode of your calculator to *a* + *bi*.

To set the mode to *a* + *bi*, follow these steps:

1. **Press [MODE] to access the mode screen.**

2. Press ⏷ repeatedly to navigate to the eighth row.

3. Press ⏵ to highlight *a* + *bi*.

4. Press ENTER to change the mode (see the second screen in Figure 5-1).

Now, try evaluating $\sqrt{-2} * \sqrt{-8}$ a second time in your calculator.

Press ⏶ to scroll through your previous calculations. When a previous entry or answer is highlighted, press ENTER to paste into your current entry line.

Success! See the result on the third screen in Figure 5-1.

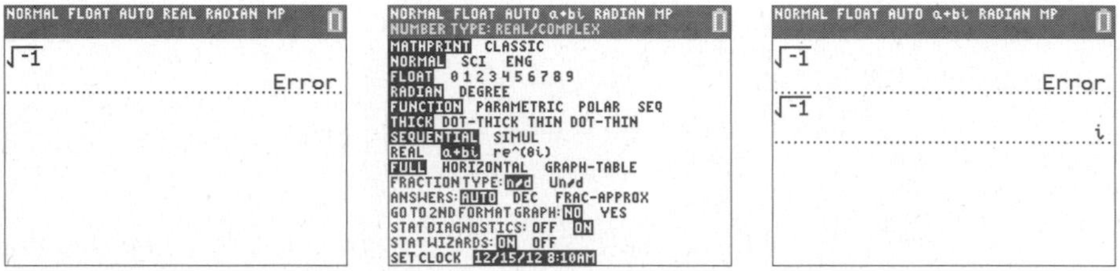

Figure 5-1:
Setting the
mode.

In Real mode   Mode screen   In *a* + *bi* mode

Entering Complex Numbers

You can enter an expression that includes the imaginary number, *i*, by pressing 2nd ⋅. Somewhere along the way, you have probably learned that $i^2 = -1$. Interestingly enough, your calculator not only knows that $i^2 = -1$, but automatically simplifies any result that would have had i^2 in it. For example, multiplying $(2 + i)(2 + i)$ would yield the trinomial, $4 + 4i + i^2$. Of course, this answer can be simplified to $3 + 4i$. Your calculator only shows the simplified answer, as shown in the first screen in Figure 5-2.

Complex numbers may not be used with the n/d fraction template. Instead, enter complex numbers as fractions using parentheses and the ÷ key. Press MATH ENTER ENTER to display the complex number answer in fraction form. See the second screen in Figure 5-2.

In *a* + *bi* mode, you can take the logarithm or square roots of negative numbers! Often, using your calculator protects you from making mistakes that are all too common for students. For example, given the expression, $\sqrt{-2} * \sqrt{-8}$, many students will mistakenly think the answer is 4. I typed the expression in the calculator and got the correct answer, –4, as shown in the third screen in Figure 5-2. Why? Before applying the order of operations, always simplify the negative inside of a square root! Here is the mathematical progression that your calculator used to simplify the given expression, $\sqrt{-2} * \sqrt{-8} = i\sqrt{2} * i\sqrt{8} = i^2\sqrt{16} = -1(4) = -4$. Pretty cool, huh?

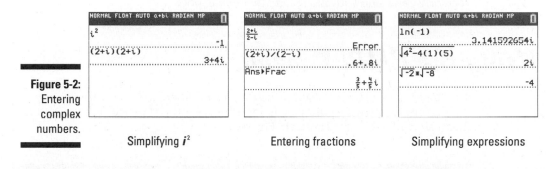

Figure 5-2:
Entering
complex
numbers.

Simplifying i^2 Entering fractions   Simplifying expressions

Interpreting Strange-looking Results

A common classroom math activity is to explore the powers of the imaginary number, *i*. Mathematics is about finding patterns, and there's an interesting pattern that emerges when you explore the powers of *i*. The results of the first four powers of *i* form a repeating pattern as *i* is raised to successive higher powers. See the first screen in Figure 5-3.

Using your calculator, something unexpected happens when you evaluate i^7. I expected the answer, –*i*. Instead, the calculator displayed -3_E-13-i, as shown in the second screen in Figure 5-3.

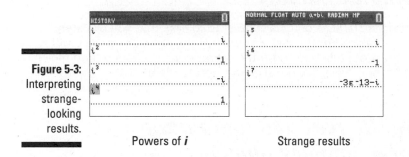

Figure 5-3:
Interpreting
strange-
looking
results.

Powers of *i*   Strange results

To decipher this strange result, you must first remember that complex numbers are written in the form $a + bi$. Using parentheses to separate the real and imaginary parts, the calculated result looks like this, $(-3_E-13) - (i)$. Now, remember that -3_E-13 is equal to $-3*10^{-13}$ in scientific notation. This is an extremely small number!

What can you learn from this strange result? You should be wary of calculated results that are extremely small! It's likely that your calculator should have returned a result of zero. The reality is that your calculator deals with approximate results all the time. You usually don't notice this because the calculator regularly comes up with the results that you expect.

Using the CMPLX menu

The functions most often used with complex numbers are all located in one convenient location on your calculator. Press [MATH]►►] to access the CMPLX menu shown in the first screen in Figure 5-4.

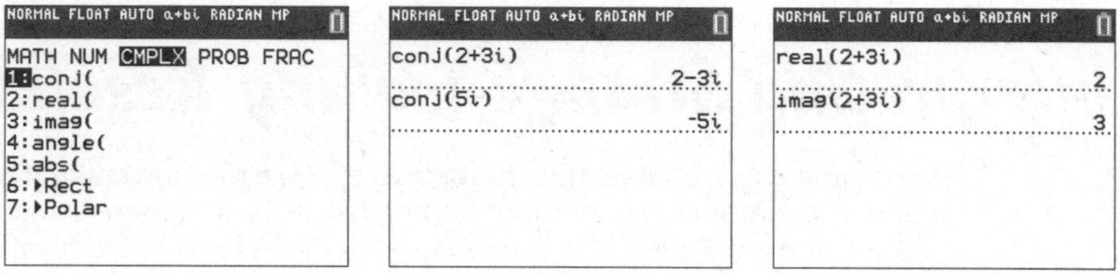

Figure 5-4: CMPLX menu functions.

CMPLX menu	Conj function	Real and Imag functions

Finding the conjugate of a complex number

Finding the conjugate of a complex number is so easy that you probably don't need a calculator for the task. In case you do, press [1] on the CMPLX menu to use the **Conj** function. Enter the expression you want to find the conjugate of and press [ENTER]. See the second screen in figure 5-4.

Finding the real and imaginary parts of a complex number

This is another function that seems to do the obvious, indicating they are mainly used in programming. In the CMPLX menu, press [2] to insert the **Real** function or press [3] to insert the **Imag** function. Enter a complex number in the argument and press ENTER to see your predictable results shown in the third screen in Figure 5-4. These tools simply identify the real or imaginary part of a complex number.

Before proceeding, press MODE and make sure your calculator is in RADIAN mode. Generally speaking, this is the recommended mode for all complex number calculations.

Finding the polar angle of a complex number

The Angle function uses the formula $\tan^{-1}(b/a)$ to calculate the polar angle of a complex number (where a is the real part and b is the imaginary part). In the CMPLX menu, press 4 to insert the Angle function, type a complex number in the argument, and then press ENTER. See the first screen in Figure 5-5.

Figure 5-5:
More
CMPLX
menu
functions.

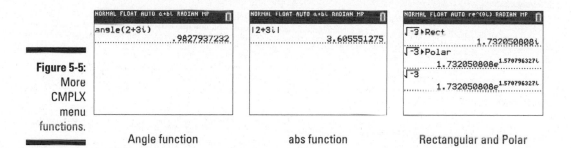

Angle function   abs function   Rectangular and Polar

Finding the magnitude (modulus) of a complex number

Entering a complex number in the absolute value template finds the magnitude (modulus) of the complex number. Algebraically your calculator uses the formula $\sqrt{(a^2 + b^2)}$ for the calculation. In the CMPLX menu, press 5 to access the **abs** function, type a complex number, and then press ENTER. See the second screen in Figure 5-5.

Press ALPHA WINDOW to access the **abs** function in the shortcut menu.

Displaying a complex result in polar or rectangular form

The last two choices in the CMPLX menu work only when inserted after typing a complex number. Use **Rect** to display a complex number in rectangular form. To display a complex number in polar form, select Polar from the CMPLX menu. See the first two lines of the last screen in Figure 5-5.

Save time converting complex numbers to polar form by changing the mode of your calculator to $re^{\wedge}(\theta\, i)$. As shown in the last line of the last screen in Figure 5-5, simply type a complex number and press ENTER to convert to polar form!

Chapter 6

Understanding the Math Menu and Submenus

A re you hunting for the absolute value function? Look no further — it's in the Math menu. Do you want to convert a decimal to a fraction? You can find the function that does this is in the Math menu as well. In general, any math function that cannot be directly accessed using the keyboard is housed in the Math menu. This chapter tells you how to access and use those functions.

Getting to Know the Math Menu and Submenus

Press MATH to access the Math menu. This menu contains five submenus: MATH, NUM, CMPLX, PROB, and FRAC. Use the ▶◀ keys to get from one submenu to the next, and back again.

The TI-84 Plus Math Menu doesn't contain the FRAC submenu. However, the FRAC menu can be accessed by pressing ALPHA Y=. If you want to learn more about the FRAC menu, see Chapter 3.

The Math MATH submenu contains the general mathematical functions such as the cubed root function (see the first screen in Figure 6-1). It also contains the calculator's Equation Solver (see the second screen in Figure 6-1) that, as you would expect, is used to solve equations. The Equation Solver is explained in Chapter 4. The Math NUM submenu contains the functions usually associated with numbers, such as the least common multiple function (see the third screen in Figure 6-1). A detailed explanation of the functions in these two menus is given later in this chapter.

The Math CMPLX submenu contains functions normally used with complex numbers. This submenu is explained in detail in Chapter 5. The Math PROB submenu contains the probability and random-number functions. (Probability is explained in Chapter 16.)

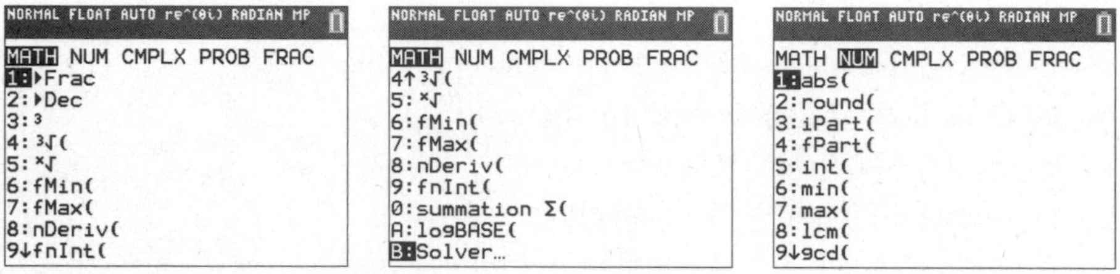

Figure 6-1:
MATH menu
and NUM
menu.

MATH submenu	MATH submenu continued	NUM submenu

Accessing Catalog Help from the Math Menu

Many of the Math menu functions have a hidden Help feature available at the press of a single key! To access the Catalog Help from the Math menu, follow these steps:

1. **Press** MATH.

2. **Use the ▷◁ keys to select the appropriate submenu of the Math menu.**

3. **Use the ▲▼ keys to place the cursor on the function you want to use.**

   I placed the cursor in front of the **fMin** function from the MATH submenu. See the first screen in Figure 6-2.

4. **Press ⊞ to access the Catalog Help.**

   I love this hidden feature! It's easy to forget the syntax for a function you don't use very often. To save time, go ahead and type the syntax directly on the Catalog Help screen before pressing ENTER. See the second screen in Figure 6-2.

5. **Press** TRACE **to PASTE or** GRAPH **to ESC.**

Two on-screen prompts in the bottom-right corner of the screen can be activated by pressing the keys located directly below the on-screen prompts. ESC takes you back to the submenu, and PASTE inserts the function as shown in the third screen in Figure 6-2.

Figure 6-2:
Accessing
Catalog
Help from
the Math
menu.

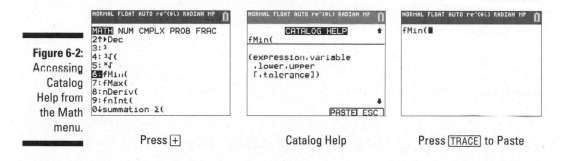

Press ➕   Catalog Help   Press TRACE to Paste

The Math MATH Submenu

Press MATH to access the Math MATH submenu. This submenu contains general mathematical functions you can insert into an expression. The following sections explain the items housed in this submenu, except for the **Solver** function at the bottom of the Math MATH submenu. This latter function, used to solve equations, is discussed in Chapter 4.

Converting between decimals and fractions

The **Frac** function always converts a finite decimal to a fraction. If your calculator can't convert a decimal to a fraction, it lets you know by redisplaying the decimal. Be sure to enter the decimal before inserting the **Frac** function. Press MATH ENTER ENTER to quickly convert a decimal answer to a fraction as shown in the first screen in Figure 6-3.

How do you convert an infinite repeating decimal into a fraction? Just type at least ten digits of the repeating decimal and press MATH ENTER ENTER. See the second screen in Figure 6-3.

The **Dec** function converts a fraction to a decimal. Enter the fraction before you insert the **Dec** function. See the third screen in Figure 6-3.

If you are in Automatic mode, include a decimal in an expression to produce a decimal answer.

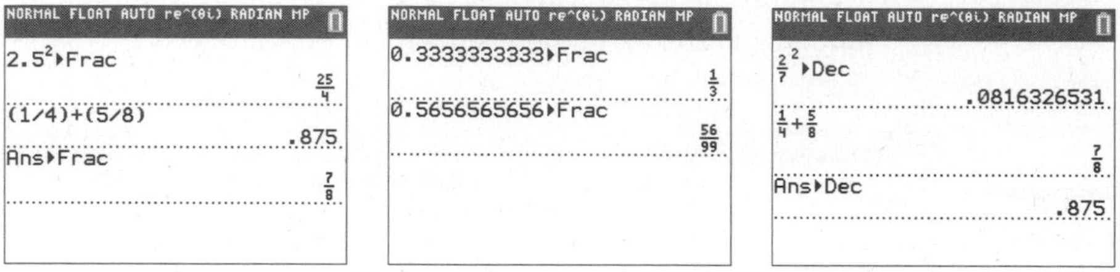

Figure 6-3:
Converting
fractions
and
decimals.

Decimals to fractions	Repeating decimals	Fractions to decimals

Cubing and taking cube roots

The cube function, ^{3}, cubes the value that precedes the function. The cube function is rarely used because it is easier to press ⌷⌷3 to cube an expression. The cube-root template, $^3\sqrt{}$, finds the cube root of a value that follows the function. See the first screen in Figure 6-4.

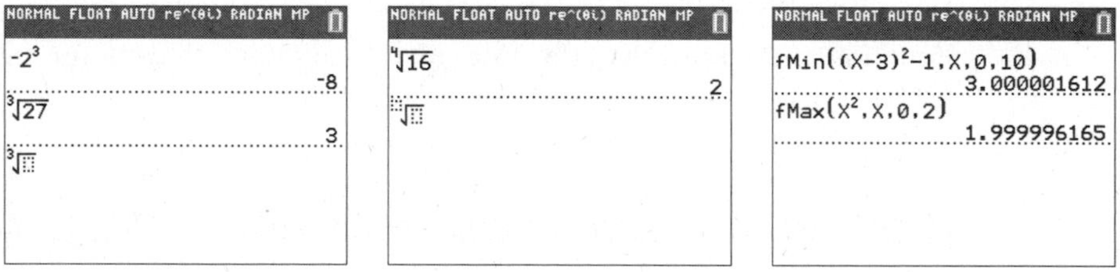

Figure 6-4:
Cubes,
roots, and
function
maximums
and
minimums.

Cube and cube root	xth root	fMin and fMax functions

Taking the xth root

The xth root template, $^x\sqrt{}$, finds the xth root of the value that follows the function. To use this function, first enter the root x, then insert the $^x\sqrt{}$ function, and then enter the argument. Alternatively, on the TI-84 Plus C you can insert the xth root template first. Then type the root, press ▶, and type the expression. This is illustrated in the second screen in Figure 6-4.

Finding the location of maximum and minimum values

The **fMin** and **fMax** functions approximate *where* the minimum or maximum value of a function occurs in a specified interval. *They do not compute the minimum or maximum value of the function;* they just give you the x-coordinate of the minimum or maximum point. Chapter 11 tells you how to get the calculator to compute minimum and maximum values of a function.

The **fMin** and **fMax** functions are stand-alone functions in the sense that they cannot be used in an expression. To use these functions, insert the appropriate function, **fMin** or **fMax**, at the beginning of a new line on the Home screen. Then enter the definition of the function whose minimum or maximum you want to locate. Press ⟨,⟩ and enter the variable used in the definition of the function you just entered. Press ⟨,⟩ and enter the lower limit of the specified interval. Then press ⟨,⟩, enter the upper limit, and press ⟨)⟩. Finally, press ⟨ENTER⟩ to *approximate* the location of the minimum or maximum in the specified interval. This is illustrated in the third screen in Figure 6-4. In this screen the calculator is *approximating* the location of the maximum value of the function x^2 in the interval $0 \le x \le 2$.

Using numerical differentiation and integration templates

The calculator cannot perform symbolic differentiation and integration. For example, it can't tell you the derivative of x^2 is $2x$, nor can it evaluate an indefinite integral. But the **nDeriv** template will approximate the derivative (slope) of a function at a specified value of the variable, and the **fnInt** template will approximate a definite integral.

Insert the **nDeriv** template. Templates are so intuitive to use that I feel silly giving you instructions. First, enter the variable you want to take the derivative with respect to and then press ⟨▶⟩. Enter the function whose derivative you want to find and then press ⟨▶⟩. Then enter the value at which the derivative is to be taken. Finally, press ⟨ENTER⟩ to *approximate* the derivative. This is illustrated in the first screen in Figure 6-5.

To use the **fnInt** template, insert **fnInt**. Enter the lower limit and press ⟨▶⟩, then enter the upper limit and press ⟨▶⟩. Enter the function you're integrating and press ⟨▶⟩. Enter the variable used in the definition of the function you just entered. Finally, press ENTER to *approximate* the definite integral. This is illustrated in the second screen in Figure 6-5.

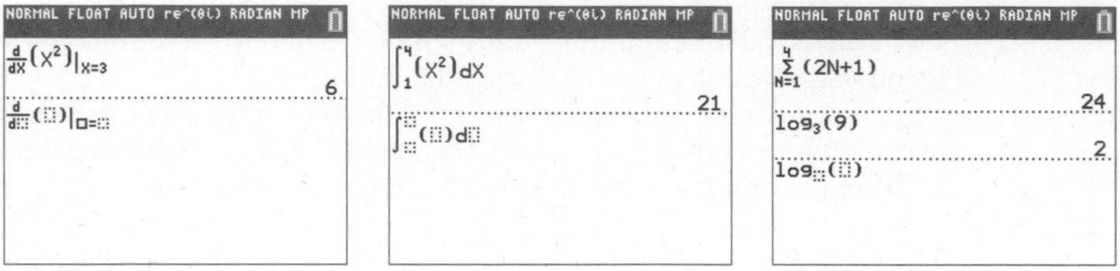

Figure 6-5:
Numerical differentiation and integration and other templates.

NORMAL FLOAT AUTO re^(θι) RADIAN MP	NORMAL FLOAT AUTO re^(θι) RADIAN MP	NORMAL FLOAT AUTO re^(θι) RADIAN MP
$\frac{d}{dx}(X^2)\big|_{X=3}$ .... 6 $\frac{d}{d\square}(\square)\big|_{\square=\square}$	$\int_{1}^{4}(X^2)dX$ .... 21 $\int_{\square}^{\square}(\square)d\square$	$\sum_{N=1}^{4}(2N+1)$ .... 24 $\log_3(9)$ .... 2 $\log_\square(\square)$

Derivative at a set point   Definitive integral   Summation and logarithms

The calculator may give you an error message or a false answer if **nDeriv** is used to find the derivative at a nondifferentiable point or if **fnInt** is used to evaluate an improper integral.

Using summation and logarithm templates

These templates can be found by pressing ▲ to scroll in the MATH menu, or by pressing ALPHA WINDOW to access the templates in the shortcut menu.

The summation template can be used to find the sum of a sequence. In math classrooms, this is commonly known as *Sigma notation*. The template should look exactly like a Sigma notation problem in your math textbook.

To use the summation template, insert **summation** Σ. Notice the cursor has a blinking "A" indicating your calculator is in Alpha mode. Press the key that corresponds to the variable you want to use and press ▶. Enter the lower limit, press ▶, then enter the upper limit and press ▶. Enter the expression and press ENTER to find the sum of the sequence as shown in the first line of the last screen in Figure 6-5.

I have good news for you! Using the logarithm template, you can change the base of a logarithm! Press MATH ▲ ▲ ENTER to insert the **logBASE** template. Simply enter the base, press ▶, and enter the number you wish to take the logarithm of. Press ENTER to display the answer. Isn't that easy and fun?

The Math NUM Submenu

Press MATH ▶ to access the Math NUM submenu. The following sections explain the items housed in the Math NUM submenu.

Finding the absolute value

The **abs** template evaluates the absolute value of the number or arithmetic expression. Insert the **abs** template, type an expression, and press ENTER. An example of using the **abs** function is illustrated in the first screen in Figure 6-6.

The **abs** template can also be found in the shortcut menu by pressing ALPHA WINDOW.

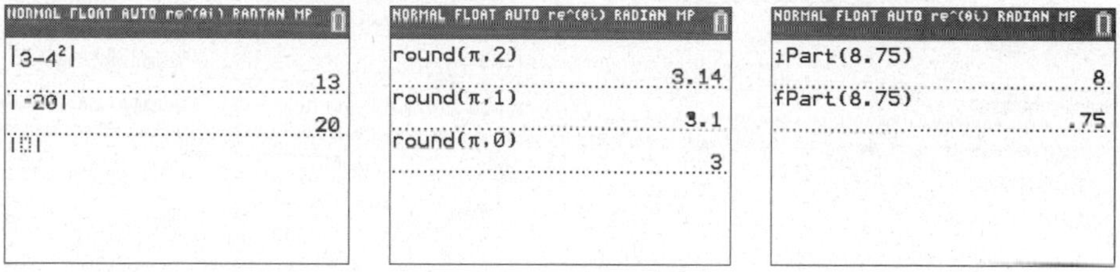

Figure 6-6:
The Math
NUM
functions.

Absolute value   Round function   iPart and fPart

Rounding numbers

The **round** function rounds a number or arithmetic expression to a specified number of decimal places. The number or expression to be rounded and the specified number of decimal places are placed after the function separated by a comma and surrounded by parentheses. The calculator supplies the opening parenthesis; you must supply the closing parenthesis. An example of using the **round** function is the second screen in Figure 6-7.

Finding the integer and fractional parts of a value

Although iPart may sound like the newest Apple product, it's actually a math function! The **iPart** and **fPart** functions (respectively) find the integer and fractional parts of the number, or the arithmetic expression that follows the function. This number or expression must be enclosed in parentheses. The calculator supplies the opening parenthesis; you must supply the closing parenthesis. An example of using the **iPart** function is the third screen in Figure 6-6.

Using the greatest-integer function

The **int** function finds the largest integer that is less than or equal to the number or arithmetic expression that follows the function. This number or expression must be enclosed in parentheses. The calculator supplies the opening parenthesis; you must supply the closing parenthesis. See the first line in the first screen in Figure 6-7.

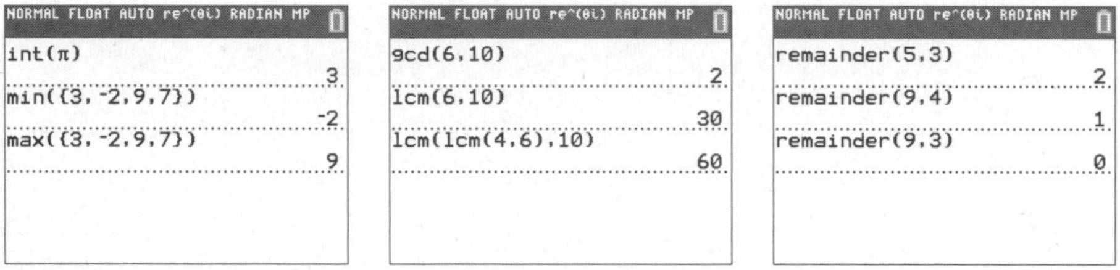

Figure 6-7:
Additional
Math NUM
functions.

NORMAL FLOAT AUTO re^(θι) RADIAN MP	
int(π)	
	3
min({3,-2,9,7})	
	-2
max({3,-2,9,7})	
	9

int, min, and max

NORMAL FLOAT AUTO re^(θι) RADIAN MP	
gcd(6,10)	
	2
lcm(6,10)	
	30
lcm(lcm(4,6),10)	
	60

lcm and gcd

NORMAL FLOAT AUTO re^(θι) RADIAN MP	
remainder(5,3)	
	2
remainder(9,4)	
	1
remainder(9,3)	
	0

remainder function

Finding minimum and maximum values in a list of numbers

The **min** and **max** functions find (respectively) the minimum and maximum values in the list of numbers that follows the function. Braces must surround the list, and commas must separate the elements in the list. You can access the braces on the calculator by pressing [2nd][(] and [2nd][)]. The list must be enclosed in parentheses. The calculator supplies the opening parenthesis; you must supply the closing parenthesis. See the last two lines in the first screen in Figure 6-7.

When using the **min** or **max** function to find the minimum or maximum of a two-element list, you can omit the braces that surround the list. For example, **min**(2, 4) returns the value 2.

Finding the least common multiple and greatest common divisor

The **lcm** and **gcd** functions find (respectively) the least common multiple and the greatest common divisor of the two numbers that follow the function. These two numbers must be separated by a comma and surrounded by parentheses. The calculator supplies the opening parenthesis; you must supply the closing parenthesis. Notice, the second screen in Figure 6-7 also demonstrates how to find the lcm of three numbers.

Finding the remainder

The **remainder** function finds the remainder resulting from dividing two positive whole numbers. The divisor cannot be zero. Press MATH ▶ 0 to insert the **remainder** function. Enter the dividend and press ⎡,⎤. Finally, enter the divisor and press ENTER. See the third screen in Figure 6-7.

Chapter 7

The Angle and Test Menus

"*W*hat mode is my calculator in?" I used to get that question all the time in my classroom. If you have the TI-84 Plus C, a quick glance at the top of the Home screen informs you of your most important mode decision: Radian or Degree mode? That's the question. It's likely that your Physics teacher needs your calculator in Degree mode and your Pre-calculus teacher wants Radian mode. How can you make everyone happy?

The reality is you need to be able to change the mode to fit the needs of the class you're in. In this chapter, I show tips on converting angles, expressions, and coordinates to the type that you need. I even show you a way to force your calculator to evaluate your angle in the correct form even if your mode isn't set correctly.

You also discover some really interesting tools hidden in the Test menu. The Test and Logic menus make it possible for you to graph piece-wise functions (graphing piece-wise functions is covered in detail in Chapter 9.)

The Angle Menu

The functions housed in the Angle menu enable you to convert between degrees and radians or convert between rectangular and polar coordinates. They also enable you to convert between decimal degrees and DMS (degrees, minutes, and seconds). You can also override the angle setting in the Mode menu of the calculator when you use these functions. For example, if the calculator is in Radian mode and you want to enter an angle measured in degrees, there's a function in the Angle menu that enables you to do so.

Converting degrees to radians

To convert degrees to radians, follow these steps:

1. **Put the calculator in Radian mode.**

   Press MODE, use the ▶◀▲▼ keys to highlight RADIAN, and then press ENTER.

2. **If necessary, press 2nd MODE to access the Home screen.**

3. **Enter the number of degrees.**

4. **Press 2nd APPS 1 to paste in the ° function.**

5. **Press ENTER to convert the degree measure to radians.**

   This is illustrated in the first screen in Figure 7-1.

   If you're a purist (like me) who likes to see radian measures expressed as a fractional multiple of π whenever possible, continuing with the following steps accomplishes this goal if it's mathematically possible.

6. **Press ÷ 2nd ^ ENTER to divide the radian measure by π.**

   This is illustrated in the second screen in Figure 7-1.

7. **Press MATH ENTER ENTER to convert the result to a fraction, if possible.**

   This is illustrated in the third screen in Figure 7-1, where 30° is equal to π/6 radians. If the calculator can't convert the decimal obtained in Step 6 to a fraction, it says so by returning the decimal in Step 7.

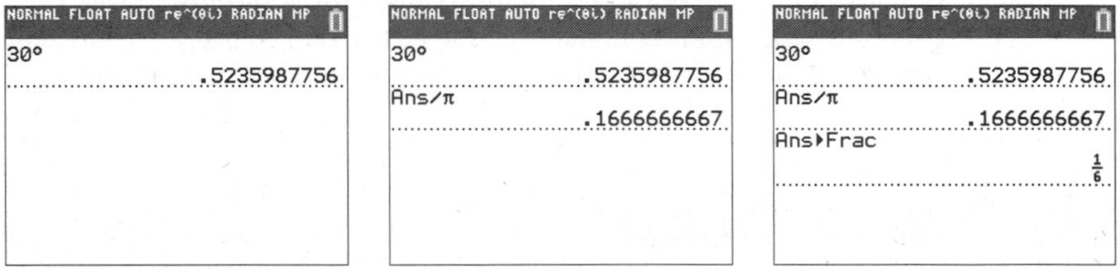

Figure 7-1: Converting between degrees and radians.

NORMAL FLOAT AUTO re^(θι) RADIAN MP	NORMAL FLOAT AUTO re^(θι) RADIAN MP	NORMAL FLOAT AUTO re^(θι) RADIAN MP
30° .5235987756	30° .5235987756 Ans/π .1666666667	30° .5235987756 Ans/π .1666666667 Ans▶Frac $\frac{1}{6}$
Degrees to radian	Divide by π	Decimal to fraction

Converting radians to degrees

To convert radians to degrees:

1. **Put the calculator in Degree mode.**

   Press MODE, use the ▶ ◀ ▲ ▼ keys to highlight DEGREE, and then press ENTER.

2. **If necessary, press 2nd MODE to access the Home screen.**

3. **Enter the radian measure.**

   If the radian measure is entered as an arithmetic expression, surround that expression with parentheses.

4. **Press 2nd APPS 3 to paste in the *r* function.**

5. **Press ENTER to convert the radian measure to degrees.**

   This is illustrated in the first screen in Figure 7-2.

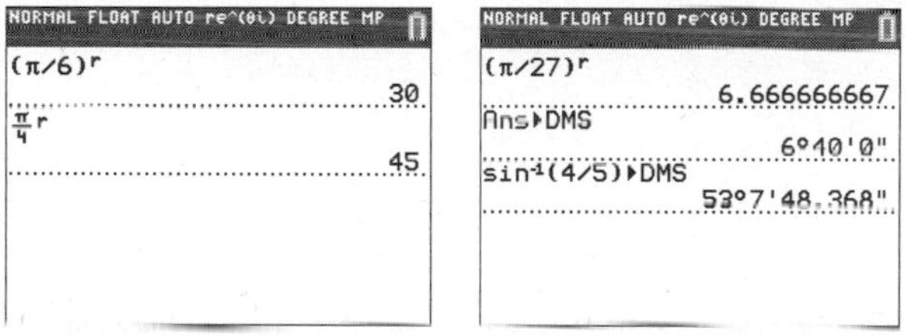

Figure 7-2:
Converting
from radian
to degrees
and from
degrees to
DMS.

Radian to degrees   Degrees to DMS

Converting between degrees and DMS

To convert decimal degrees to DMS (degrees, minutes, and seconds), follow these steps:

1. **Put the calculator in Degree mode.**

   Press MODE, use the ▶ ◀ ▲ ▼ keys to highlight DEGREE, and then press ENTER.

2. **If necessary, press 2nd MODE to access the Home screen.**

3. **Enter the degree measure.**

4. **Press 2nd APPS 4 ENTER to convert the degrees to DMS.**

   This is illustrated in the second screen in Figure 7-2.

Entering angles in DMS measure

To enter an angle in DMS measure (and convert to decimal degrees), follow these steps:

1. **Enter the number of degrees and press** 2nd APPS 1 **to insert the degree symbol.**

2. **Enter the number of minutes and press** 2nd APPS 2 **to insert the symbol for minutes.**

3. **Enter the number of seconds and press** ALPHA + **to insert the symbol for seconds.**

4. **Press** ENTER **to evaluate your DMS measure.**

   Since your calculator is in Degree, pressing ENTER converts DMS to decimal degrees. See the first screen in Figure 7-3.

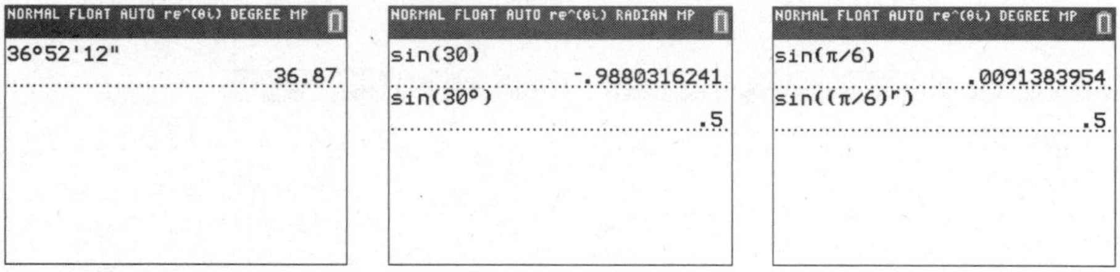

NORMAL FLOAT AUTO re^(θι) DEGREE MP	NORMAL FLOAT AUTO re^(θι) RADIAN MP	NORMAL FLOAT AUTO re^(θι) DEGREE MP
36°52'12" 36.87	sin(30) -.9880316241 sin(30°) .5	sin(π/6) .0091383954 sin((π/6)ʳ) .5
Entering DMS	Using degree symbol	Using radian-measure symbol

Figure 7-3: Entering DMS and overriding the mode.

Overriding the mode of the angle

If the calculator is in Radian mode but you want to enter an angle in degrees, enter the number of degrees and then press 2nd APPS 1 to insert the ° degree symbol. Essentially, you're forcing your calculator to evaluate your angle in degrees regardless of the mode setting. Getting into the habit of adding the degree symbol to your angle gives your math teacher the warm fuzzies all over!

If the calculator is in Degree mode and you want to enter an angle in radian measure, enter the number of radians and then press 2nd APPS 3 to insert the radian-measure symbol.

If the radian measure is entered as an arithmetical expression, such as $\pi/4$, be sure to surround it with parentheses before you insert the radian-measure symbol!

Converting rectangular and polar coordinates

All four of the conversion tools discussed in this section operate the same way. You must insert the conversion tool you desire, then enter the coordinates and press [ENTER].

Before you start your converting fun, decide the mode you want your calculator to be in. I chose Radian mode for these calculations. Press [MODE], use [▼] to highlight RADIAN, and then press [ENTER] to put your calculator in Radian mode.

Here are two easy-to-use tools that convert rectangular coordinates to polar coordinates:

- ✔ **R▶Pr:** This tool converts rectangular coordinates to polar coordinates and produces the *r* value.

- ✔ **R▶Pθ:** This tool converts rectangular coordinates to polar coordinates and produces the θ value. See the first screen in Figure 7-4.

What if you want to do the conversion from polar to rectangular coordinates? No worries! These tools can help you out:

- ✔ **P▶Rx:** This tool converts polar coordinates to rectangular coordinates and produces the *x* value.

- ✔ **P▶Ry:** This tool converts polar coordinates to rectangular coordinates and produces the *y* value. See the second screen in Figure 7-4.

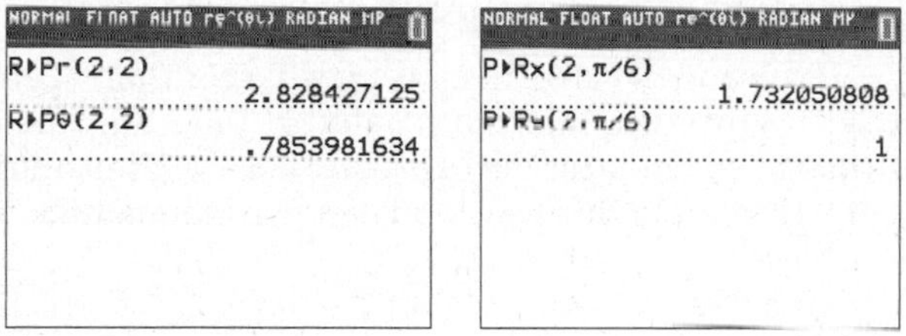

Figure 7-4: Converting rectangular and polar coordinates.

Rectangular to polar   Polar to rectangular

The Test Menu

The often overlooked Test menu enables you to use your calculator in creative ways to solve problems. Do you want to do better on your next standardized test? Keep reading because some of these tips just might help your score.

Understanding Boolean logic

Have you ever wondered how your calculator "thinks"? Your calculator employs Boolean logic and prefers to work with the integers 1 and 0. These are called truth values with 1 meaning True and 0 meaning False. Boolean operators like **and**, **or**, and **not** help your calculator organize its "thoughts."

How does this help you? Just remember this: 1=True and 0=False.

Comparing numbers

The Test menu has a list of relational operators that you can use to compare values. To access the Test menu, press [2nd][MATH].

Follow these steps to compare two expressions:

1. **Enter an expression.**

   If necessary, press [2nd][MODE] to access the Home screen. See the first screen in Figure 7-5.

2. **Press [2nd][MATH] to access the Test menu.**

   See the second screen in Figure 7-5.

3. **Press the number associated with the relational operator you want.**

4. **Enter an expression.**

   This is illustrated in the third screen in Figure 7-5.

5. **Press [ENTER] to evaluate the comparison statement.**

   Remember, 1=True and 0=False. Because 6 is larger than 3, entering the expression 5+1 > 3 produced the truth value of 1 (True). Similarly, 6 is not smaller than 3, so entering the expression 5 + 1 < 3 produced the truth value of 0 (False). In this way, you can use inequalities to compare two values.

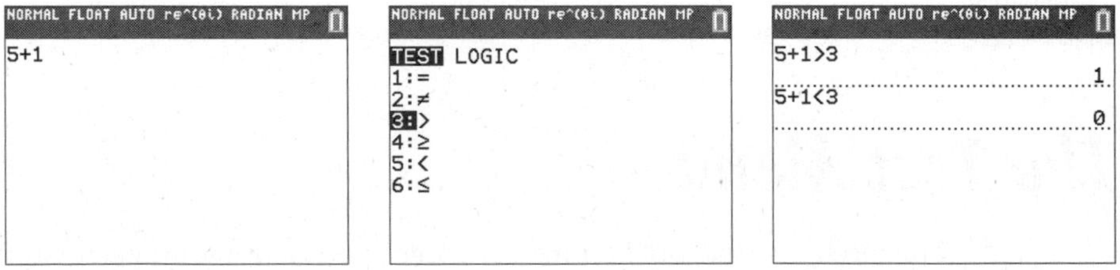

Figure 7-5:
Comparing
numbers.

NORMAL FLOAT AUTO re^(θi) RADIAN MP	NORMAL FLOAT AUTO re^(θi) RADIAN MP	NORMAL FLOAT AUTO re^(θi) RADIAN MP

5+1	**TEST** LOGIC
	1: =
	2: ≠
	3: >
	4: ≥
	5: <
	6: ≤
		5+1>3
		                    1
		5+1<3
		                    0

Enter expression	Test menu	1=True, 0=False

Testing equivalent expressions

Does $2 + 2 = 4$? You can check and see by entering the equation in your calculator. Try it! Press 2nd MATH ENTER to type an = sign. Of course, you'll probably want to use this feature for slightly more complicated problems. See the first screen in Figure 7-6.

Have you ever seen a question like this on a standardized test?

Evaluate $\log_2(5) + \log_2(3)$

A) $\log_2(5\text{-}3)$ C) $\log_2(5*3)$

B) $\log_2(5+3)$ D) $\log_2(5/3)$

As long as you can enter a logarithm in your calculator (MATH ▲ ▲ ENTER), you can solve this problem. Enter the expression, insert the = sign from the Test menu, and then enter one of the answers. When your calculator returns 1 (True), you have found the correct answer. See the second screen in Figure 7-6.

TIP

Press ▲ to scroll through your previous calculations. When a previous entry or answer is highlighted, press ENTER to paste into your current entry line (where you can edit the expression.)

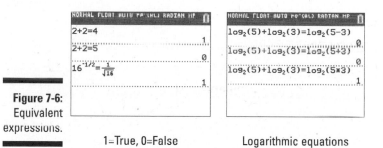

Figure 7-6:
Equivalent
expressions.

1=True, 0=False   Logarithmic equations

Using Logic commands

Compound inequalities are two inequalities that are joined by the word **"and"** or by the word **"or."** Often, an **"and"** inequality is written in a shortcut form where two inequalities are sandwiched together. For example, $2 < x < 5$ can also be written as a compound inequality: $2 < x$ and $x < 5$. Sorry, that last statement makes me uncomfortable. I like this statement better: $x > 2$ and $x < 5$. Thanks! Now, I'll be able to sleep better tonight.

Press 2nd MATH ▶ to access the Logic menu. See the first screen in Figure 7-7.

For an **"and"** compound inequality to be true, both statements must be true. See the second screen in Figure 7-6. For an **" or"** compound inequality to be true, only one of the statements must be true. See the third screen in Figure 7-7.

Two more commands are also found in the Logic menu: **xor** and **not**. These are used almost exclusively in programming. The command **xor** means exactly one statement is true. The **not** command flips everything that is true to false, and vice versa.

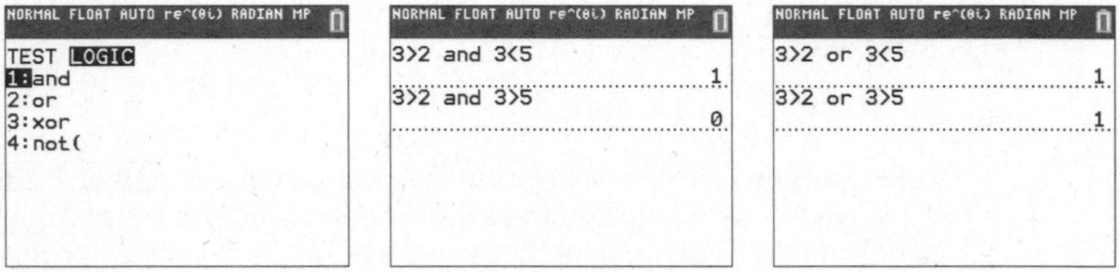

Figure 7-7: Logic commands.

NORMAL FLOAT AUTO re^(θi) RADIAN MP	NORMAL FLOAT AUTO re^(θi) RADIAN MP	NORMAL FLOAT AUTO re^(θi) RADIAN MP
TEST **LOGIC**	3>2 and 3<5	3>2 or 3<5
1:and	1	1
2:or	3>2 and 3>5	3>2 or 3>5
3:xor	0	1
4:not(		
Logic menu	and statements	or statements

Chapter 8

Creating and Editing Matrices

In This Chapter

▶ Entering and editing a matrix

▶ Storing a matrix

▶ Doing matrix arithmetic

▶ Finding determinants and other matrix operations

▶ Solving systems of equations using matrices

▶ Converting a matrix to reduced row-cchelon form

A matrix is a rectangular array of numbers arranged in rows and columns. The dimensions, r × c, of a matrix are defined by the number of rows and columns in the matrix. The individual elements in a matrix are called *elements*. Why is it that students are more familiar with the movie, *The Matrix,* rather than the actual mathematics? But I digress . . .

What are matrices used for? There are several scientific applications, but in the math classroom, they are mainly used to solve systems of equations. In this chapter, you learn the basics of dealing with matrices. There are many rules associated with matrix operations; if you break one, you should expect an error message. Keep reading so you can avoid error messages altogether!

Entering Matrices

Using the hidden MTRX shortcut menu is my preferred method of entering matrices (the easy way). Alternatively, you can use the Matrix editor found by pressing [2nd][x⁻¹] (the hard way). Here are the instructions for entering matrices the easy way:

1. **Press [ALPHA][ZOOM] to display the Quick Matrix Editor.**

   I like the name! Quick and easy!

2. **Use the** ▶◀▲▼ **keys to highlight the dimensions you want and press** [ENTER].

The default dimensions of a matrix are two rows by two columns. I created a 3×2 matrix. See the first screen Figure 8-1.

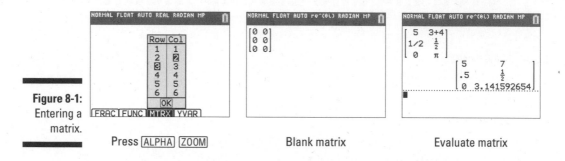

Figure 8-1:
Entering a
matrix.

Press [ALPHA] [ZOOM]   Blank matrix   Evaluate matrix

3. **Use** ▼ **to highlight the word OK and then press** [ENTER].

See the result in the second screen in Figure 8-1.

4. **Enter an expression and press** ▶ **to advance to the next element in the matrix.**

Repeat this step until you have filled in every element in the matrix. When you press ▶ after entering the last element in the first row, the calculator moves to the beginning of the second row and waits for you to make another entry.

To enter a fraction, delete the zero first and then press [ALPHA][Y=] to use the n/d fraction template.

5. **Press** [ENTER] **to evaluate the matrix.**

This is illustrated in the third screen in Figure 8-1.

You cannot copy and paste a matrix output from the calculator history. This is not a deal breaker! You may copy and paste the matrix expression you entered as many times as you would like.

Storing a Matrix

Storing a matrix is a handy feature to have around. This is especially helpful if you're reusing the same matrix in future calculations. Follow these steps to store a matrix:

1. **Enter a matrix on the Home screen.**

   See the preceding section for details — but don't press ENTER yet!

2. **Position your cursor to the right of your matrix and press** STO▶.

   See the first screen in Figure 8-2.

3. **Press** 2nd x⁻¹ **and then press** ENTER **(to choose Matrix [A]).**

   Welcome to the Matrix editor. See the second screen in Figure 8-2.

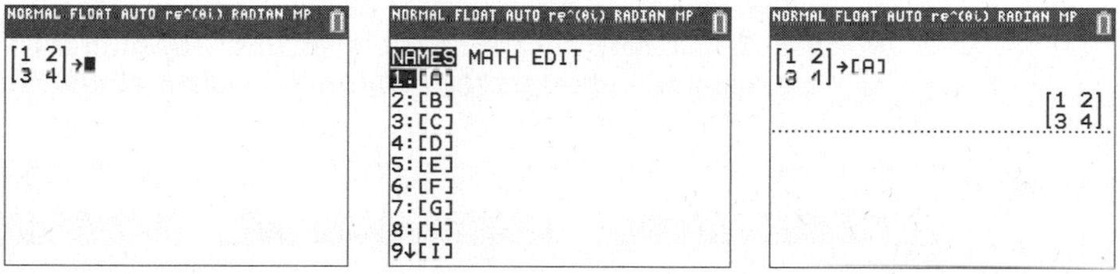

NORMAL FLOAT AUTO re^(θi) RADIAN MP	NORMAL FLOAT AUTO re^(θi) RADIAN MP	NORMAL FLOAT AUTO re^(θi) RADIAN MP
$\begin{bmatrix} 1 & 2 \\ 3 & 4 \end{bmatrix}$→■	**NAMES** MATH EDIT **1:[A]** 2:[B] 3:[C] 4:[D] 5:[E] 6:[F] 7:[G] 8:[H] 9↓[I]	$\begin{bmatrix} 1 & 2 \\ 3 & 4 \end{bmatrix}$→[A] $\begin{bmatrix} 1 & 2 \\ 3 & 4 \end{bmatrix}$

Figure 8-2:
Storing a
matrix.

      Press STO▶   Matrix editor   Press ENTER

4. **Finally, press** ENTER.

   See the third screen in Figure 8-2.

You can delete a stored matrix by pressing 2nd + 2 5 and then pressing DEL with your cursor on the matrix you want to delete. However, deleting a matrix is completely unnecessary. Saving a different matrix as [A] overwrites the current matrix [A].

Matrix Arithmetic

When evaluating arithmetic expressions that involve matrices, you usually want to perform the following basic operations: scalar multiplication, addition, subtraction, and multiplication. You might also want to raise a matrix to an integral power.

Be careful! Matrix arithmetic is not like the arithmetic you've been doing for years. Expect the unexpected! Multiplying two matrices is quite different than multiplying two numbers.

Here's how you enter matrix operations in an arithmetic expression:

1. **Enter a matrix on the Home screen.**

   To paste the name of a matrix into an expression, press ⟦2nd⟧⟦x^{-1}⟧ and key in the number of the matrix name. I chose to use matrix [A]. Alternatively, you can press ⟦ALPHA⟧⟦ZOOM⟧ to quickly create a new matrix.

2. **Enter the operations you want to perform and press ⟦ENTER⟧ when you're finished.**

   Here's how you enter the various operations into the arithmetic expression:

   - **Entering the scalar multiple of a matrix:** To enter the scalar multiple of a matrix in an arithmetic expression, enter the value of the scalar and then enter the name of the matrix, as shown in the first screen in Figure 8-3.

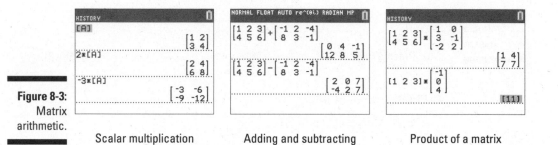

Figure 8-3:
Matrix
arithmetic.

Scalar multiplication   Adding and subtracting   Product of a matrix

- **Adding or subtracting matrices:** When adding or subtracting matrices, both matrices must have the same dimensions. If they don't, you get the ERROR: DIMENSION MISMATCH error message.

  Entering the addition and subtraction of matrices is straightforward; just combine the matrices by pressing ⟦+⟧ or ⟦−⟧, as appropriate. The second screen in Figure 8-3 illustrates this process.

- **Multiplying two matrices:** When finding the product A*B of two matrices, the number of columns in the first matrix (matrix A) must equal the number or rows in the second matrix (matrix B). If this condition isn't satisfied, you get the ERROR: DIMENSION MISMATCH error message.

  Matrix multiplication is a tricky process. However, entering matrix multiplication in a calculator is straightforward; just multiply the matrices by pressing ⟦×⟧, as shown in the third screen in Figure 8-3.

- **Raising a matrix to a positive integral power:** When finding the power of a matrix, the matrix must be *square* (number of rows = number of columns). If it isn't, you get the ERROR: INVALID DIMENSION error message.

Only non-negative integers can be used for the power of a matrix. If the exponent is a negative integer, you get the ERROR: DOMAIN error message.

Look at the top of the first screen in Figure 8-4. Is that the answer you expect to get when you square a matrix? It's better to think of squaring a matrix as multiplying a matrix by itself, as shown at the bottom of the first screen in Figure 8-4.

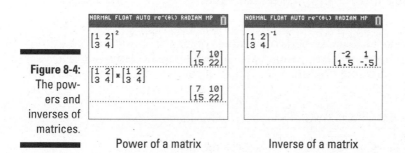

Figure 8-4: The powers and inverses of matrices.

Power of a matrix   Inverse of a matrix

- **Finding the inverse of a matrix:** When finding the inverse of a matrix, the matrix must be *square* (number of rows = number of columns) and *nonsingular* (nonzero determinant). If it isn't square, you get the ERROR: INVALID DIMENSION error message. If it is singular (determinant = 0), you get the ERROR: SINGULAR MATRIX error message. Evaluating the determinant of a matrix is explained in the next section.

  Enter the inverse of a matrix by entering the matrix and then pressing $\boxed{x^{-1}}$, as shown in the second screen of Figure 8-4.

It may look like you're putting a matrix to the power of –1 when your press $\boxed{x^{-1}}$. That isn't the case! In this setting, $[A]^{-1}$ is read as "the inverse of matrix A" or "inverting matrix [A]." This is similar to the notation that's used for inverse functions.

Evaluating the Determinant and Other Matrix Operations

Quite a few operations are unique to matrices. All the matrix-specific operations are found by pressing $\boxed{2nd}\boxed{x^{-1}}\boxed{\blacktriangleright}$. This is called the MATRX MATH Operations menu (see the first two screens in Figure 8-5). I'm not going to go through every command in this list, but I do explain some of the most popular matrix operations.

MATRX MATH   More MATRX MATH   Determinant

The determinant is used to perform all kinds of matrix operations, so the determinant is a good place to start. When finding the determinant of a matrix, the matrix must be square (number of rows = number of columns). If it isn't, you get the ERROR: INVALID DIMENSION error message.

To evaluate the determinant of a matrix, follow these steps:

1. **If necessary, press** 2nd MODE **to access the Home screen.**

2. **Press** 2nd x^{-1} ▶ 1 **to select the det(command from the MATRX MATH menu.**

3. **Enter the matrix.**

   Press ALPHA ZOOM to create a matrix from scratch, or press 2nd x^{-1} to access a stored matrix.

4. **Press** ENTER **to evaluate the determinant.**

   This procedure is illustrated in the third screen in Figure 8-5.

There are a few other skills that you will need when working with matrices. These skills are easily done by hand, but if you already have the matrix typed in your calculator, why not let the calculator do the work for you to save time?

- **Transposing of a matrix:** To transpose a matrix, enter the matrix and then press 2nd x^{-1} ▶ 2 to select the Transpose command from the MATRX MATH menu. See the first screen in Figure 8-6.

- **Entering the identity matrix:** You don't have to enter a matrix in order to find the identity matrix. To enter an identity matrix in an expression, press 2nd x^{-1} ▶ 5 to select the identity command from the MATRX MATH menu. Then enter the size of the identity matrix. For example, enter 2 for the 2×2 identity matrix, as shown in the second screen in Figure 8-6.

If you raise a square matrix to the zero power, you get the identity matrix. See the third screen in Figure 8-6. How cool is that?

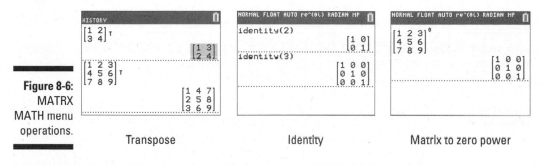

Figure 8-6:
MATRX
MATH menu
operations.

Transpose   Identity   Matrix to zero power

Solving a System of Equations

Finally, we get to the good stuff! Matrices are the perfect tool for solving systems of equations (the larger the better). All you need to do is decide which method you want to use.

$A^{-1} * B$ method

What do the A and B represent? The letters A and B are capitalized because they refer to matrices. Specifically, A is the coefficient matrix and B is the constant matrix. In addition, X is the variable matrix. No matter which method you use, it's important to be able to convert back and forth from a system of equations (shown below) to matrix form shown in Figure 8-7.

$$2x + 3y - 2z = 8$$
$$x \quad\;\; - 4z = 1$$
$$2x - y - 6z = 4$$

Here's a short explanation of where this method comes from. Any system of equations can be written as the matrix equation, $A * X = B$. By pre-multiplying each side of the equation by A^{-1} and simplifying, you get the equation $X = A^{-1} * B$.

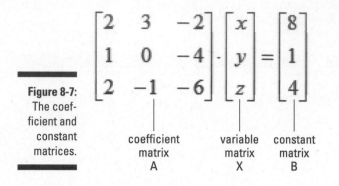

Figure 8-7:
The coefficient and constant matrices.

$$
\underbrace{\begin{bmatrix} 2 & 3 & -2 \\ 1 & 0 & -4 \\ 2 & -1 & -6 \end{bmatrix}}_{\substack{\text{coefficient} \\ \text{matrix} \\ \text{A}}} \cdot \underbrace{\begin{bmatrix} x \\ y \\ z \end{bmatrix}}_{\substack{\text{variable} \\ \text{matrix} \\ \text{X}}} = \underbrace{\begin{bmatrix} 8 \\ 1 \\ 4 \end{bmatrix}}_{\substack{\text{constant} \\ \text{matrix} \\ \text{B}}}
$$

Using your calculator to find $A^{-1} * B$ is a piece of cake. Just follow these steps:

1. **Enter the coefficient matrix, A.**

   Press ALPHA ZOOM to create a matrix from scratch or press 2nd x^{-1} to access a stored matrix. See the first screen in Figure 8-8.

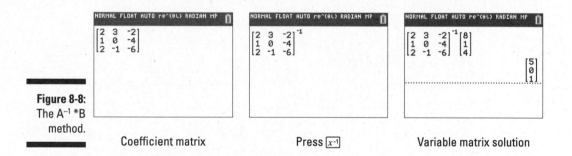

Figure 8-8:
The $A^{-1} * B$ method.

Coefficient matrix   Press x^{-1} Variable matrix solution

2. **Press x^{-1} to find the inverse of matrix A.**

   See the second screen in Figure 8-8.

3. **Enter the constant matrix, B.**

4. **Press ENTER to evaluate the variable matrix, X.**

   The variable matrix indicates the solutions: $x=5$, $y=0$, and $z=1$. See the third screen in Figure 8-8.

If the determinant of matrix A is zero, you get the ERROR: SINGULAR MATRIX error message. This means that the system of equations has either no solution or infinite solutions.

Augmenting matrices method

Augmenting two matrices enables you to append one matrix to another matrix. Both matrices must be defined and have the same number of rows. Use the system of equations (shown below) to augment the coefficient matrix and the constant matrix.

$$2x + 3y - 2z = 8$$
$$x \quad - 4z = 1$$
$$2x - y - 6z = 4$$

To augment two matrices, follow these steps:

1. **Press 2nd x⁻¹ ▶ 7 to select the Augment command from the MATRX MATH menu.**

2. **Enter the first matrix and then press ⬚ (see the first screen in Figure 8-9).**

   To create a matrix from scratch, press ALPHA ZOOM. To access a stored matrix, press 2nd x⁻¹.

3. **Enter the second matrix and then press ENTER.**

   The second screen in Figure 8-9 displays the augmented matrix.

4. **Store your augmented matrix by pressing STO▶ 2nd x⁻¹ 3 ENTER.**

   I stored the augmented matrix as [C]. See the third screen in Figure 8-9.

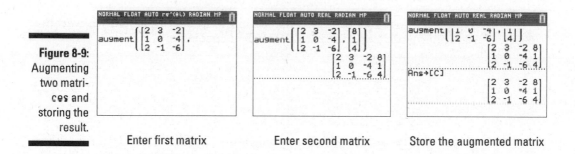

Figure 8-9: Augmenting two matrices and storing the result.

Enter first matrix   Enter second matrix   Store the augmented matrix

Systems of linear equations can be solved by first putting the augmented matrix for the system in reduced row-echelon form. The mathematical definition of reduced row-echelon form isn't important here. It's simply an equivalent form of the original system of equations, which, when converted back to a system of equations, gives you the solutions (if any) to the original system of equations.

To find the reduced row-echelon form of a matrix, follow these steps:

1. **Press** 2nd x⁻¹ ▶ **and use** ▲ **to scroll to the rref(function in the MATRX MATH menu.**

   See the first screen in Figure 8-10.

Figure 8-10:
Finding the
reduced
row-
echelon
form.

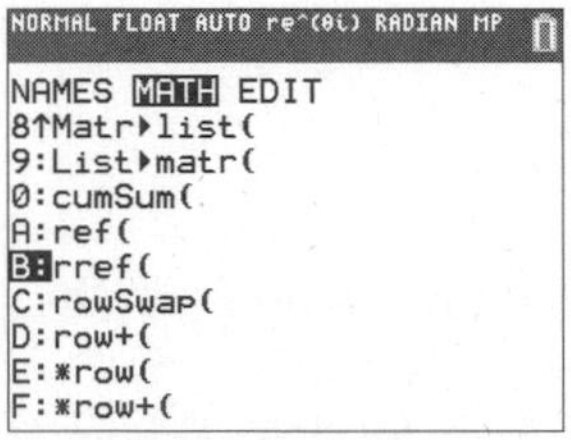

NORMAL FLOAT AUTO re^(θL) RADIAN MP

NAMES **MATH** EDIT
8↑Matr▶list(
9:List▶matr(
0:cumSum(
A:ref(
B:rref(
C:rowSwap(
D:row+(
E:*row(
F:*row+(

rref(function

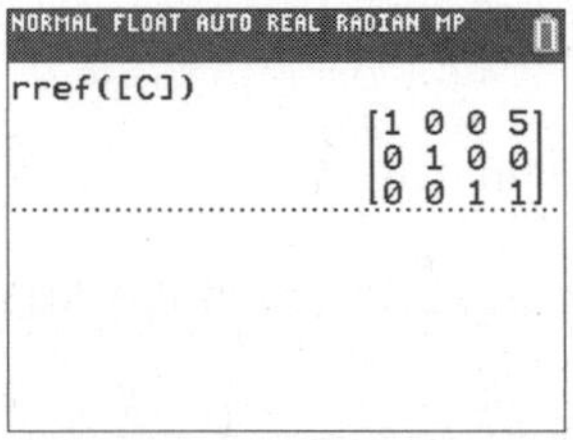

NORMAL FLOAT AUTO REAL RADIAN MP

rref([C])

$$\begin{bmatrix} 1 & 0 & 0 & 5 \\ 0 & 1 & 0 & 0 \\ 0 & 0 & 1 & 1 \end{bmatrix}$$

Solution matrix

2. **Press** ENTER **to paste the function on the Home screen.**

3. **Press** 2nd x⁻¹ **and press** 3 **to choose the augmented matrix you just stored.**

4. **Press** ENTER **to find the solution.**

   See the second screen in Figure 8-10.

To find the solutions (if any) to the original system of equations, convert the reduced row-echelon matrix to a system of equations:

$$1x + 0y + 0z = 5$$
$$0x + 1y + 0z = 0$$
$$0x + 0y + 1z = 1$$

As you see, the solutions to the system are $x=5$, $y=0$, and $z=1$. Unfortunately, not all systems of equations have unique solutions like this system. Here are examples of the two other cases that you may see when solving systems of equations:

$2x - y = 3$	$1x + 6y = 10$
$2x - y = -4$	$-2x - 12y = -20$
(no solution)	(infinite solutions)

See the reduced row-echelon matrix solutions to the preceding systems in the first two screens in Figure 8-11. To find the solutions (if any), I convert the reduced row-echelon matrices to a system of equations:

$1x - 0.5y = 0$ $1x + 6y = 10$

$0x + 0y = 1$ $0x + 0y = 0$

(no solution)   (infinite solutions)

Because one of the equations in the first system simplifies to 0=1, this system has no solution. In the second system, one of the equations simplifies to 0=0. This indicates the system has an infinite number of solutions that are on the line x+6y–10.

Figure 8-11:
Solving
systems of
equations
that don't
have unique
solutions.

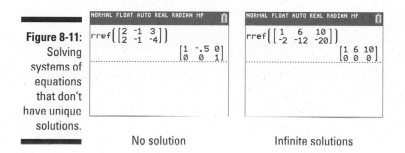

No solution   Infinite solutions

Part III
Graphing and Analyzing Functions

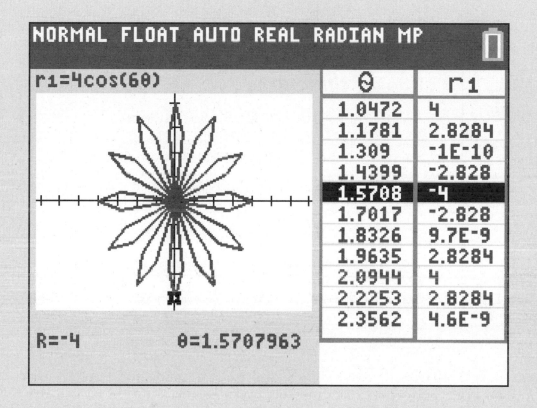

Check out the video for enabling testing mode using the TI-TestGuard app and (more importantly) disabling testing mode at www.dummies.com/extras/ti84plus.

In this part . . .

- ✔ Learn the basics of graphing a function, setting the window, and changing the color and graph style of a function.

- ✔ Make use of the Trace and Zoom features to explore graphed functions.

- ✔ Find out how to evaluate a function and find its critical points.

- ✔ Employ the Inequality Graphing app to graph and analyze inequalities.

- ✔ Get instructions for graphing and evaluating parametric and polar equations.

- ✔ See how to enter and graph explicit and recursive sequences.

Chapter 9

Graphing Functions

....

In This Chapter

▶ Entering functions into your calculator

▶ Making graph formatting settings

▶ Graphing functions

▶ Changing the color and style of your graph

▶ Graphing families of functions

▶ Graphing piecewise and trigonometric functions

▶ Viewing the graph and the function on the same screen

▶ Saving and recalling a graph

....

The calculator has a variety of features that help you easily graph a function. The first step is to enter the function into the calculator. Then to graph the function, you set the viewing window and press GRAPH. You might want to change the color of the function you graph (there are 15 colors to choose from). But why stop there? Why not change the color of your axes and graph border while you're at it? If you like the way graph paper looks, you could consider adding gridlines to your graph as well. And if you're graphing trig functions, you may want to customize the window to improve the look and functionality of your graph.

As you can see, there are a lot of choices to be made to get the graph to look exactly the way you want it to look! Keep reading to find out the details and hopefully learn a few new things along the way.

Entering Functions

Before you can graph a function, you must enter it into the calculator. The calculator can handle up to ten functions at once, Y_1 through Y_9 and Y_0. To enter functions in the calculator, perform the following steps:

1. **Press** MODE **and put the calculator in Function mode.**

   To highlight an item in the Mode menu, use the ▶◀▲▼ keys to place the cursor on the item and then press ENTER. Highlight **FUNCTION** in the fourth line to put the calculator in Function mode. See the first screen in Figure 9-1.

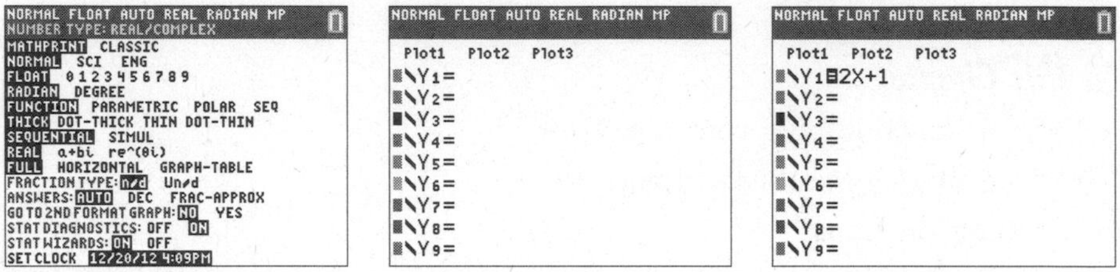

Figure 9-1:
Setting the mode and entering functions.

Function mode   Y= editor   Entering functions

2. **Press** Y= **to access the Y= editor.**

   See the second screen in Figure 9-1.

3. **Enter your function.**

   If necessary, press CLEAR to erase a previous function entry. Then enter your function.

Your math textbook may use a function notation like this: f(x)=2x+1. To graph a function in your calculator, you must realize **f(x)** is interchangeable with **y**, only the notation differs. See the third screen in Figure 9-1.

When you're defining functions, the only symbol the calculator allows for the independent variable is the letter X. Press X,T,θ,n to enter this letter in your function.

As a timesaver, when entering functions in the Y= editor, you can reference another function. Use the shortcut Y-VAR menu to paste a function name in the function you're entering in the Y= editor. Just press ALPHA TRACE and choose the name of the function you want to insert in your equation. See the first screen in Figure 9-2.

How does calling up the name of another function save you time? Well, say you're trying to graph a circle in your calculator with the equation $x^2 + y^2 = 36$. Of course, you need to solve the equation for y to graph the circle equation in your calculator. Solving for y gives you: $y = \pm\sqrt{\left(36 - x^2\right)}$. Notice, it takes two functions to graph a circle! No problem. In function Y_1 I enter $Y_1 = \sqrt{\left(36 - X^2\right)}$. Then, to save time, I use the shortcut Y–VAR menu to enter $Y_2 = -Y_1$. See the second screen in Figure 9-2.

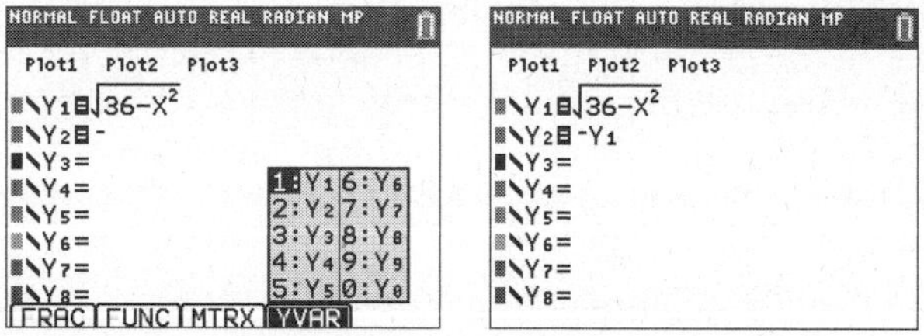

Figure 9-2:
Referencing
another
function
in the
Y= editor.

Shortcut menu   Insert Y₁

Formatting Your Graph

Set the graph format settings by following these steps:

1. **Press** 2nd ZOOM **to access the Format menu.**

   See the first screen in Figure 9-3. Alternatively, access the Format menu from the Mode menu; press MODE, and then highlight YES on the line that says, GO TO 2ND FORMAT GRAPH.

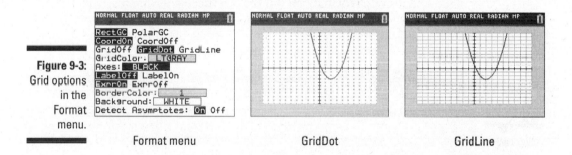

Figure 9-3:
Grid options
in the
Format
menu.

Format menu   GridDot   GridLine

2. **Set the format for the graph by using the** ▶ ◀ ▲ ▼ **keys to place the cursor on the desired format, and then press** ENTER **to highlight it.**

   In the Format menu, each line of the menu will have one item highlighted. Keep in mind, the TI-84 Plus does not have as many options in the Format menu as the TI-84 Plus C does. An explanation of each menu selection follows:

   • **RectGC and PolarGC:** This gives you a choice between having the coordinates of the location of the cursor displayed in (x, y) rectangular form or in (r, θ) polar form. Select **RectGC** for rectangular form or **PolarGC** for polar form.

- **GridOff, GridDot, and GridLine:** If you select **GridDot**, grid points appear in the graph at the intersections of the tick marks on the *x*- and *y*-axes. See the second screen in Figure 9-3. If you select **GridOff**, no grid points appear in the graph. If you select **GridLine**, your graph background looks a lot like graph paper, as shown in the third screen in Figure 9-3.

- **GridColor:** If you place your cursor on **GridColor**, a menu spinner is activated. Use the ▶◀ keys to choose one of 15 grid colors. The default color is light gray.

- **Axes:** If you place your cursor on **Axes**, a spinner is activated. Use the ▶◀ keys to choose one of 15 Axes colors. Axes Off is also a choice in the spinner. See the first screen in Figure 9-4. The default Axes color is black.

- **LabelOff and LabelOn:** If you want the *x*- and *y*-axes to be labeled, select **LabelOn** (as in the second screen in Figure 9-4). Because the location of the labels isn't ideal, selecting **LabelOff** is usually a wise choice.

- **CoordOn and CoordOff:** This tells the calculator whether you want to see the coordinates of the cursor location displayed in the Graph border at the bottom of the screen as you move the cursor. Select **CoordOn** if you want to see these coordinates; select **CoordOff** if you don't. I always keep CoordOn. See the third screen in Figure 9-4.

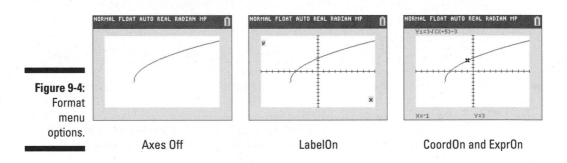

Figure 9-4:
Format
menu
options.

Axes Off   LabelOn   CoordOn and ExprOn

- **ExprOn and ExprOff:** If you select **ExprOn**, when you're tracing the graph of a function, the definition of that function appears in the Graph border in the upper-left corner of the screen (see the third screen in Figure 9-4). If you select **ExprOff** and **CoordOn**, then only the number of the function appears when you trace the function. If you select **ExprOff** and **CoordOff**, then nothing at all appears on the screen to indicate which function you're tracing.

- **BorderColor:** If you place your cursor on **BorderColor**, a spinner is activated. Use the ▶◀ keys to choose one of four border colors. Why only four color choices? Many of the other colors would be too dark, making it difficult to read the expression and coordinate information in the Graph border. The four color choices are: 1-Light Gray (default), 2-Light Green, 3-Teal, and 4-White. See the first screen in Figure 9-5.

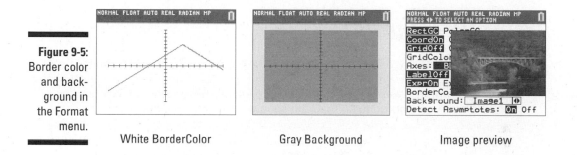

Figure 9-5: Border color and background in the Format menu.

White BorderColor   Gray Background   Image preview

- **Background:** If you place your cursor on **Background**, a spinner is activated. Use the ▶◀ keys to choose one of 15 colors that can serve as the background of your graph page. See the second screen in Figure 9-5. In addition to colors, you can use the spinner to choose one of 10 images (see Chapter 22 for more info on inserting images). A preview of each image automatically pops up as you scroll through the spinner. See the third screen in Figure 9-5. The default background is set to Off.

- **Detect Asymptotes:** If you select **Detect Asymptotes On**, vertical asymptotes will not have any points graphed where the vertical asymptote is located as shown in the first screen in Figure 9-6. Another way of thinking about this is your calculator is not trying to connect every point graphed to the next (across singularities). If you select **Detect Asymptotes Off**, the graph rate increases. This means there's a strong likelihood that there will be a vertical line where the vertical asymptote is located as illustrated in the second screen in Figure 9-6. Confused? Your calculator is trying to connect all the points that are graphed. So, if the limit of the function is positive and negative infinity on opposite sides of the vertical asymptote, a vertical line will appear because your calculator is trying to connect the points on each side of the vertical asymptote.

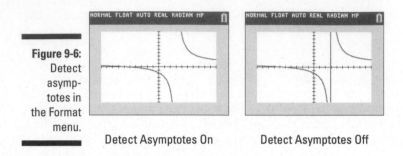

Figure 9-6:
Detect
asymp-
totes in
the Format
menu.

Detect Asymptotes On   Detect Asymptotes Off

Graphing Functions

After you have entered the functions into the calculator and formatted your graph, you're almost ready to start your graphing fun. Once you get the hang of graphing, you won't need to go through all these steps. Right now, I'm being very thorough so that you soon will be graphing like a pro!

Turning off Stat Plots (if necessary)

The top line in the Y= editor tells you the graphing status of the Stat Plots. (Stat Plots are discussed in Chapter 18.) If **Plot1**, **Plot2**, or **Plot3** is highlighted, then that Stat Plot will be graphed along with the graph of your functions. If it's not highlighted, it won't be graphed. In the first screen in Figure 9-7, **Plot1** is highlighted and will be graphed along with the functions in the Y= editor.

To turn off a highlighted Stat Plot in the Y= editor, use the ▶◀▲▼ keys to place the cursor on the highlighted Stat Plot and then press [ENTER]. See the second screen in Figure 9-7. The same process is used to highlight the Stat Plot again in order to graph it at a later time.

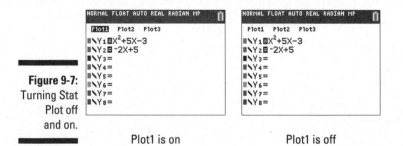

Figure 9-7:
Turning Stat
Plot off
and on.

Plot1 is on   Plot1 is off

When you're graphing functions, Stat Plots can be a nuisance if they're turned on when you don't really want them to be graphed. The most common symptom of this problem is the ERROR: INVALID DIMENSION error message — which by itself gives you almost no insight into what's causing the problem. So if you aren't planning to graph a Stat Plot along with your functions, make sure all Stat Plots are turned off!

Selecting and deselecting a function

I remember the first time I saw this happen: A student in my class had the correct function entered in Y_1 but the function wasn't showing up on the graph. I couldn't figure out why this was happening — you have to be very observant to catch the problem (no pun intended)! Do you see the difference between the two screens in Figure 9-8? It turns out my student had accidentally turned off Y_1 by pressing ENTER with the cursor on the equal sign.

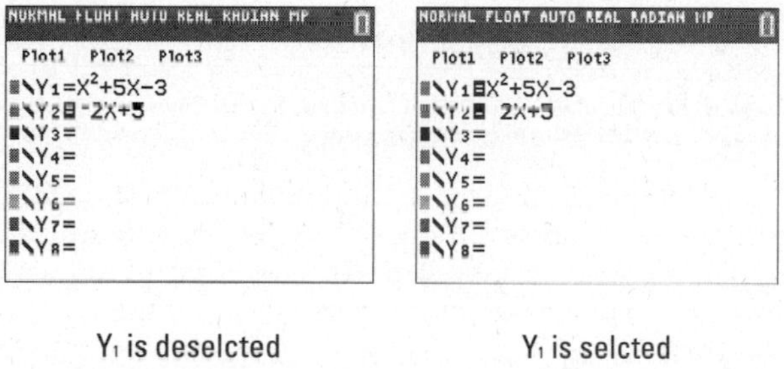

Figure 9-8:
Select (turn on) and Deselect (turn off) a function.

NORMAL FLOAT AUTO REAL RADIAN MP

Plot1 Plot2 Plot3
■\Y₁=X²+5X−3
■\Y₂= -2X+5
■\Y₃=
■\Y₄=
■\Y₅=
■\Y₆=
■\Y₇=
■\Y₈=

Y_1 is deselcted

NORMAL FLOAT AUTO REAL RADIAN MP

Plot1 Plot2 Plot3
■\Y₁=X²+5X−3
■\Y₂= 2X+5
■\Y₃=
■\Y₄=
■\Y₅=
■\Y₆=
■\Y₇=
■\Y₈=

Y_1 is selcted

Deselect (turn off) Y_1 and Y_2 by removing the highlight from their equal signs. This is done in the Y= editor by using the ▶ ◀ ▲ ▼ keys to place the cursor on the equal sign and then pressing ENTER to toggle the equal sign between highlighted and not highlighted. The calculator graphs a function only when its equal sign is highlighted!

Adjusting the graph window

When you graph a function, you usually can't see the whole graph. You are limited to viewing the graphing window, which typically shows only a small portion of the function. There are four values that determine the portion of the coordinate plane you can see: Xmin, Xmax, Ymin, and Ymax. Press WINDOW to display the current window variables. See the two screens in Figure 9-9.

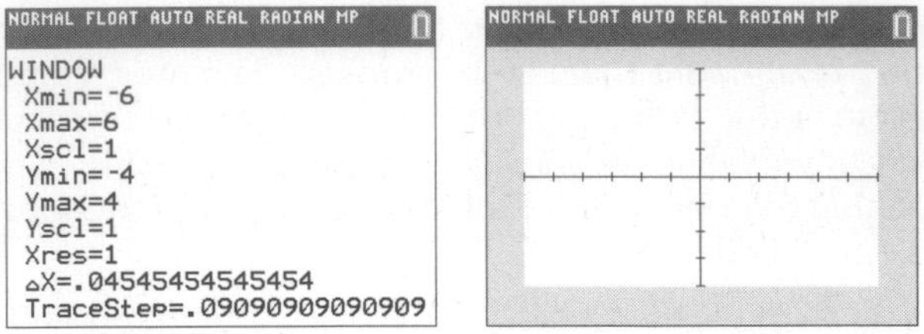

NORMAL FLOAT AUTO REAL RADIAN MP

WINDOW
 Xmin=-6
 Xmax=6
 Xscl=1
 Ymin=-4
 Ymax=4
 Yscl=1
 Xres=1
 △X=.04545454545454
 TraceStep=.09090909090909

NORMAL FLOAT AUTO REAL RADIAN MP

Figure 9-9:
Graphing
window.

Window editor   Window variables

It takes practice to find a good viewing window for the function you're graphing. Here are the steps needed to set the window of your graph:

1. **Press WINDOW to access the Window editor.**

2. **After each of the window variables, enter a numerical value that is appropriate for the functions you're graphing. Press ENTER after entering each number.**

Entering a new window value automatically clears the old value.

Make sure your (Xmin < Xmax) and (Ymin < Ymax) or you'll get the ERROR: WINDOW RANGE error message.

Editing your Window variables is a good place to start as you search for a good viewing window. In addition, using the Zoom features described in Chapter 10 may be necessary to perfect your graphing window. The following gives an explanation of the variables you must set to adjust the graphing window:

- **Xmin and Xmax:** These are, respectively, the smallest and largest values of *x* in view on the *x*-axis.

If you don't know what values your graph will need for **Xmin** and **Xmax**, press ZOOM 6 to invoke the **ZStandard** command. This command automatically graphs your functions in the Standard viewing window.

- **Xscl:** This is the distance between tick marks on the *x*-axis. (Go easy on the tick marks; using too many makes the axis look like a railroad track. Twenty or fewer tick marks makes for a nice looking *x*-axis.)

If you want to turn off tick marks altogether, set **Xscl=0** and **Yscl=0**.

- **Ymin and Ymax:** These are, respectively, the smallest and largest values of *y* that will be placed on the *y*-axis.

If you have assigned values to **Xmin** and **Xmax** but don't know what values to assign to **Ymin** and **Ymax**, press ZOOM 0 to invoke the **ZoomFit** command. This command uses the **Xmin** and **Xmax**

settings to determine the appropriate settings for **Ymin** and **Ymax**, and then automatically draws the graph.

- **Yscl:** This is the distance between tick marks on the *y*-axis. (As with the *x*-axis, too many tick marks make the axis look like a railroad track. Fifteen or fewer tick marks is a nice number for the *y*-axis.)

- **Xres:** This setting determines the resolution of the graph. It can be set to any of the integers 1 through 8. When **Xres** set equal to 1, the calculator evaluates the function at each of the 133 pixels on the *x*-axis and graphs the result. If **Xres** is set equal to 8, the function is evaluated and graphed at every eighth pixel.

   Xres is usually set equal to 1. If you're graphing a lot of functions, it may take the calculator a while to graph them at this resolution. If you change **Xres** to a higher number, your function will graph quicker, but you may not get as accurate of a graph.

- **ΔX and TraceStep:** These two variables are linked together, and **TraceStep** is always twice as big as **ΔX** value. **ΔX** determines how your cursor moves on a graph screen in "free trace." **TraceStep** controls the X-value jump when you are tracing a function on a graph screen. For more on Tracing a graph, see Chapter 10.

3. **Press** GRAPH **to graph the functions.**

Stopping or pausing your graph

After pressing GRAPH, there's usually a small delay before you begin to see your function plotting on the graph from left to right. If it's taking a long time for the calculator to graph your functions (maybe your **Xres** setting is too small), press ON to terminate the graphing process. I also love having the capability to pause your graph! Simply press ENTER to pause the plotting of your graph and then press ENTER again to resume graphing. See the two screens in Figure 9-10. Notice, the elliptical busy indicator in the top right corner of the screen indicating that your calculator is working hard.

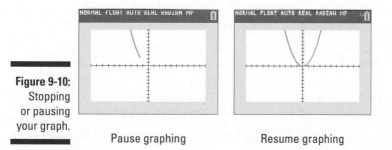

Figure 9-10:
Stopping
or pausing
your graph.

Pause graphing   Resume graphing

Adjusting Your Color/Line Settings

If you're graphing several functions at once, your calculator automatically graphs each function in a different color. If you have the TI-84 Plus C, you may want to change the colors more to your liking or further distinguish your functions by choosing a different graph style. To do this, follow these steps:

1. **Press** Y= **to access the Y= editor.**

2. **Use the** ◄ **key to place the cursor on the two-piece icon appearing to the left of the equal sign.**

   See the first screen in Figure 9-11. The icon displays two pieces, a rectangular color indicator and a line style icon. I love taking a quick glance to the left of a function and identifying the color and graph line style.

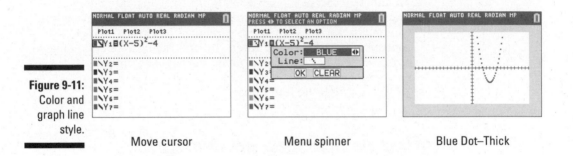

Figure 9-11: Color and graph line style.

Move cursor   Menu spinner   Blue Dot–Thick

3. **Press** ENTER **to open the Color / Line selection menu.**

   See the second screen in Figure 9-11.

4. **Use the** ►◄ **keys to operate the spinner menu until you get the desired graph color.**

   There are 15 colors to choose from: Blue, Red, Black, Magenta, Green, Orange, Brown, Navy, Light Blue, Yellow, White, Light Gray, Medium Gray, Gray, and Dark Gray.

5. **Use the** ▼ **key or press** ENTER **to navigate your cursor to the next selection field in the menu.**

6. **Use the** ►◄ **keys to operate the spinner menu until you get the desired graph line style.**

   You have eight graph line styles to choose from: ╲ (Line), ╿ (Thick Line–Default), ╗ (shading above the curve), ╚ (shading below the curve), ⊣ (Path), ◊ (Animate), ╲ (Dotted–Thick), and ╲ (Dotted–Thin).

7. **Use the** ▼ **key or press** ENTER **to navigate your cursor to OK, and press** ENTER.

If you change your mind, navigate to CLEAR and press ENTER. This nulli-fies any changes you made to the color or graph line style. See an exam-ple of a function with dotted–thick graph line style in the third screen in Figure 9-11.

Here are the different line styles available:

- **Thin Line, Thick Line, Dotted Thin Line, and Dotted Thick Line:** The default graph style setting is Thick Line. See the first screen in Figure 9-12.

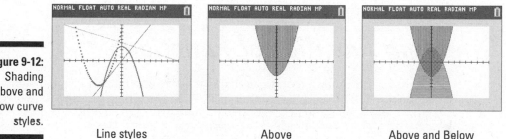

Figure 9-12: Shading above and below curve styles.

Line styles   Above   Above and Below

- **Shading above and below the curve styles:** See the second screen in Figure 9-12 for an example of shading above the curve style.

  Your calculator has four shading patterns: vertical lines, horizon-tal lines, negatively sloping diagonal lines, and positively sloping diagonal lines. These patterns help you to distinguish the solu-tion region for a system of inequalities. See the third screen in Figure 9-12.

- **Path and Animated styles:** The Path style, denoted by the ⊕ icon, uses a circle to indicate a point as it's being graphed (as illustrated in the first screen in Figure 9-13). When the graph is complete, the circle disappears and leaves the graph in Line style.

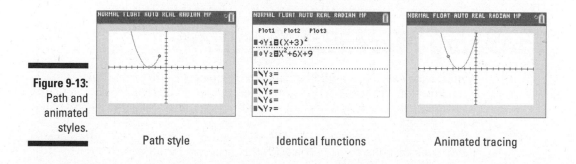

Figure 9-13: Path and animated styles.

Path style   Identical functions   Animated tracing

The animate style, denoted by the ⬚ icon, also uses a circle to indicate a point as it's being graphed, but when the graph is complete, no graph appears on the screen. For example, if this style is used, graphing $y = -x^2 + 9$ looks like a movie of the path of a ball thrown in the air.

I like to use the animate tool to show students that two functions are identical. At first glance, the functions in the second screen in Figure 9-13 may not look identical. However, the animate tool shows they are identical as the bubble traces the original function. See the third screen in Figure 9-13.

Graphing Families of Functions

There's a little known technique that can be used to quickly graph a family of functions. The secret is to enter a list as an element of an expression. You must enter a list using brackets {} with numbers separated by commas.

For this example, I want to simulate the family of functions represented by the function: $f(x) = x^2 + c$, where c is an integer. In order to graph a few examples from the family of functions, I graph $Y_1 = x^2 + \{-3, -1, 1, 3\}$. See the screens in Figure 9-14.

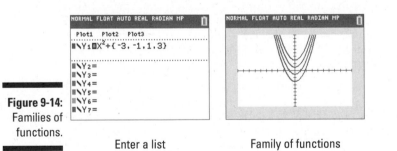

Figure 9-14: Families of functions.

Enter a list   Family of functions

Graphing Piecewise Functions

A *piecewise function* is actually made up of "pieces" of different functions. Each function "piece" is defined over a certain interval. Using your calculator to graph piecewise functions can be a bit tricky, but you'll get the hang of it soon enough.

Before I explain the steps involved in this process, I want to explain why it works. Your calculator evaluates statements and produces one of two possible truth values: 1 = *True* and 0 = *False*. When is the statement $x < -2$ true?

Your calculator returns a 1 (for True) when x is smaller than –2, and returns a 0 (for False) when x is equal to or larger than –2. If you divide a function by the statement $x < -2$, the function is divided by 1 when the statement is true, and the function is divided by 0 when the statement is false. Guess what? You can't divide by 0, so the function will not graph on the interval where the statement is false!

In order to graph the piecewise function below, I start by graphing each piece as a separate function on the calculator. Then I tell you how to graph all three pieces in one function so you can show off to your friends.

$$f(x) = \begin{cases} x+8 & x < -1 \\ x^2 & -1 \le x \le 2 \\ 1-x & x > 2 \end{cases}$$

Here are the steps to graph a piecewise function in your calculator:

1. **Press** ALPHA Y= ENTER **to insert the n/d fraction template in the Y= editor.**

2. **Enter the function piece in the numerator and enter the corresponding interval in the denominator.**

   To enter the first function piece in Y_1, I enter $(X + 8)$ in the numerator and $(X < -1)$ in the denominator.

   Press 2nd MATH to insert an inequality from the Test menu. Press 2nd MATH ▶ to insert "**and**" from the Logic menu.

   Your calculator can't evaluate a sandwiched inequality like this one: $(-1 < X < 2)$. Fortunately, $(-1 < X < 2)$ can also be written as a compound inequality: $(-1 \le X)$ and $(X \le 2)$. See the first screen in Figure 9-15.

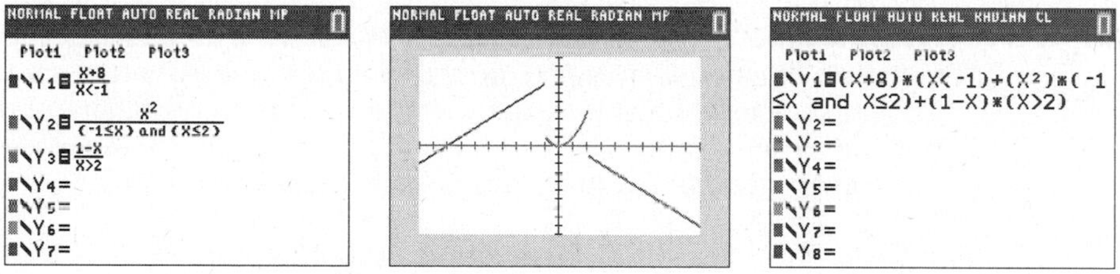

Figure 9-15:
Graphing
piecewise
functions.

Using division   Piecewise graph   Using multiplication

3. **Press** GRAPH **to graph the function pieces.**

   See the second screen in Figure 9-15.

If you can successfully graph piecewise functions in your calculator, you're well on your way to becoming addicted to your calculator. If you know two different methods to graph piecewise functions in your calculator, then you may need to enter a 12-step program for calculator addiction! The first step is admitting you have a problem. In case you're wondering, I think I may have a problem!

Here's a method of graphing piecewise functions all in one function:

1. **In the Y= editor, enter the first function piece using parentheses and multiply by the corresponding interval (also in parentheses).**

   Don't press ENTER yet!

2. **Press ⊞ after each piece and repeat until finished.**

   Refer to the third screen in Figure 9-15. In order for you to see the whole equation, I temporarily switched the calculator to Classic mode.

There are a few advantages to graphing piecewise functions all in one function. Obviously, only one function in the Y= editor is used. Another advantage is when using Trace, you don't have to jump from function to function (see Chapter 10). One disadvantage shows up if a piecewise function has a domain other than all Real numbers. When multiplication is used, a horizontal line at $y = 0$ graphs for any missing domain intervals.

 If one or more of the functions in your piecewise-defined function is a trigonometric function, make sure the calculator is in Radian and not Degree mode. Otherwise, your piecewise-defined function may look like a step function instead of the graph you were expecting. The next section tells you how to change the mode and how to graph trigonometric functions.

Graphing Trig Functions

The calculator has built-in features especially designed for graphing trigonometric functions. They produce graphs that look like graphs you see in textbooks, and when you trace these graphs, the x-coordinate of the tracing point is always given as a fractional multiple of π. To use these features when graphing trigonometric functions, follow these steps:

1. **Put the calculator in Function and Radian mode.**

   Press MODE. In the fourth line, highlight **Radian**, and in the fifth line highlight **Function**. (To highlight an item in the Mode menu, use the ▶◀▲▼ keys to place the cursor on the item, and then press ENTER.)

2. **Enter your trigonometric functions into the Y= editor.**

   See the first screen in Figure 9-16.

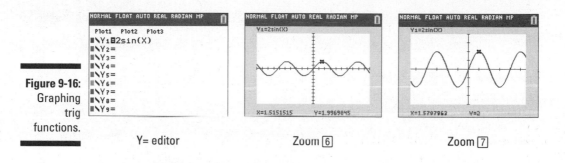

Figure 9-16:
Graphing
trig
functions.

Y= editor   Zoom [6]   Zoom [7]

3. **Press** [ZOOM][7] **to graph the function.**

[ZOOM][7] invokes the **ZTrig** command that graphs the function in a viewing window in which $-66\pi/24 \le x \le 66\pi/24$ and $-4 \le y \le 4$. It also sets the tick marks on the x-axis to multiples of $\pi/2$. Compare the graphing windows of ZStandard ([ZOOM][6]) and ZTrig in the last two screens in Figure 9-16. I like the ZTrig window better. What do you think?

When you trace a function graphed in a **ZTrig** window, the x-coordinate of the trace cursor will be a multiple of $\pi/24$, although the x-coordinate displayed at the bottom of the screen will be a decimal approximation of this value. (Tracing is explained in Chapter 10.)

If you want to graph trigonometric functions in Degree mode, press [ZOOM][7]. The ZTrig window automatically adjusts to account for the mode of your calculator. Isn't that nice?

Viewing the Function and Graph on the Same Screen

If you're planning to play around with the definition of a function you're graphing, it's quite handy to have both the Y= editor and the graph on the same screen. That way you can edit the definition of your function and see the effect your editing has on your graph. To do so, follow these steps:

1. **Put the calculator in Horizontal mode.**

Press [MODE] and highlight **Horizontal** in the ninth line of the menu, as illustrated in the first screen in Figure 9-17. To highlight an item in the Mode menu, use the [▷][◁][▲][▽] keys to place the cursor on the item, and then press [ENTER].

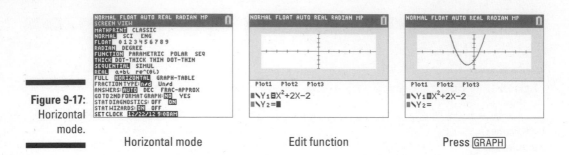

Figure 9-17:
Horizontal
mode.

Horizontal mode   Edit function   Press GRAPH

You may notice that the graphing window dimensions have not changed, but the graph now only takes up half of the screen. Having a smaller graphing window is the price you pay for viewing the Y= editor on the same screen. If you don't like the look of the graph, try pressing ZOOM 5 to square the graphing window.

2. **Press Y=.**

   The Graph window appears at the top of the screen and the Y= editor at the bottom of the screen.

3. **Enter or edit a function in the Y= editor.**

   See the second screen in Figure 9-17.

4. **Press GRAPH to graph the function.**

   The graph doesn't update after entering a function in the Y= editor. You must press GRAPH to update the graphing screen. See the third screen in Figure 9-17.

To edit or enter a function, press Y=. To see the resulting graph, press GRAPH.

Saving and Recalling a Graph

If you want to save your graph as a Graph Database, when you recall the graph at a later time, the graph remains interactive. This means that you can, for example, trace the graph and resize the viewing window because a Graph Database also saves the Graph Mode, Window, Format, and Y= editor settings. It does not, however, save the split-screen settings (**Horizontal** and **Graph-Table**) entered in the ninth line of the Mode menu. This section explains how to save, delete, and recall a graph in a Graph Database.

To save a Graph Database, perform the following steps:

1. **Press** [2nd][PRGM][▶][▶] **to access the Draw STO menu.**

   See the first screen in Figure 9-18.

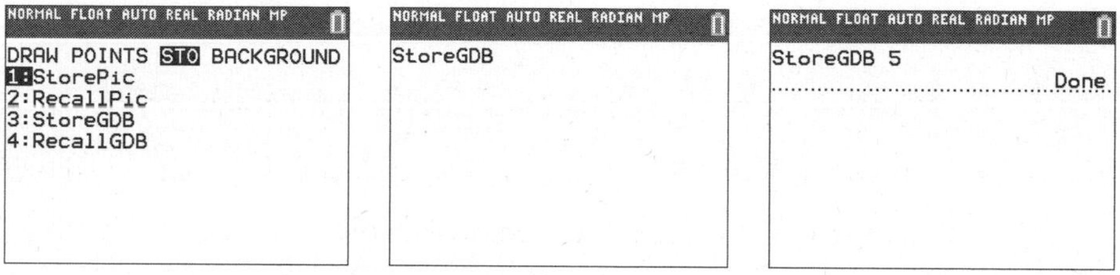

NORMAL FLOAT AUTO REAL RADIAN MP

DRAW POINTS **STO** BACKGROUND
1:StorePic
2:RecallPic
3:StoreGDB
4:RecallGDB

NORMAL FLOAT AUTO REAL RADIAN MP

StoreGDB

NORMAL FLOAT AUTO REAL RADIAN MP

StoreGDB 5
....Done

Figure 9-18:
Saving
in Graph
Database.

STO menu   Store GDB   Press [ENTER]

2. **Press** [3] **to store your graph as a Graph Database.**

   See the second screen in Figure 9-18.

3. **Enter an integer 0 through 9.**

   The calculator can store up to ten Graph Databases. If, for example, you enter the number **5**, your Graph Database is stored in the calculator as **GDB 5**.

   If you save your Graph Database as **GDB 5** without realizing that you had previously stored another Graph Database as **GDB 5**, the calculator — without warning or asking your permission — erases the old **GDB 5** and replaces it with the new **GDB 5**. To see a list of the Graph Databases already stored in your calculator, press [2nd][+][2][9].

   If you already have ten Graph Databases stored in your calculator and don't want to sacrifice any of them, consider saving some of them on your PC. Chapter 20 describes how to do this.

4. **Press** [ENTER].

   See the third screen in Figure 9-18.

To delete a Graph Database from your calculator, perform the following steps:

1. **Press** [2nd][+] **to access the Memory menu.**

2. **Press** [2] **to access the Mem ManagementDelete menu.**

3. **Press** [9] **to access the GBD files stored in the calculator.**

4. **If necessary, repeatedly press** [▼] **to move the indicator to the GBD you want to delete.**

5. **Press** [DEL].

If there is more than one Graph Database stored in your calculator, you are asked whether or not you really want to delete this item. Press ☐2☐ if you want it deleted, or press ☐1☐ if you have changed your mind about deleting it.

6. **Press** ☐2nd☐☐MODE☐ **to exit this menu and return to the Home screen.**

To recall a saved Graph Database, perform the following steps:

1. **Press** ☐2nd☐☐PRGM☐☐▶☐☐▶☐ **to access the Draw STO menu.**

2. **Press** ☐4☐ **to recall your Graph Database.**

3. **Enter the number of your stored Graph Database.**

4. **Press** ☐ENTER☐.

Chapter 10

Exploring Functions

· ·

In This Chapter

▶ Tracing the graph of a function

▶ Using Zoom commands

▶ Constructing tables of functional values

▶ Creating and clearing user-defined tables

▶ Viewing graphs and tables on the same screen

· ·

The calculator has three very useful features that help you explore the graph of a function: tracing, zooming, and creating tables of functional values. Tracing shows you the coordinates of the points that make up the graph. Zooming enables you to quickly adjust the viewing window for the graph so you can get a better idea of the nature of the graph. And creating a table — well, I'm sure you already know what that shows you. This chapter explains how to use each of these features.

The TI-84 Plus C has a graph border on the edge of the graph screen where the calculator displays functions, trace values, and helpful hints. The TI-84 Plus displays most of the same information, but it does so directly on the bottom of the graph screen. Additionally, the TI-84 Plus C has better screen resolution, so there are some small differences in the zoom window settings mentioned in this chapter if you use the TI-84 Plus calculator.

Tracing a Graph

After you graph your function (described in the Chapter 9), you can press TRACE and use ▶ and ◀ to more closely investigate the function.

If you use only the ▶◀▲▼ keys (called a *free-moving trace*) instead of TRACE to locate a point on a graph, all you will get is an *approximation* of the location of that point; you rarely get an actual point on the graph. So always use TRACE to identify points on a graph!

The following list describes what you see, or don't see, as you trace a graph:

✔ **The definition of the function:** The function you're tracing is displayed in the top border of the screen, provided the calculator is in **ExprOn** format (refer to Chapter 9). If the Format menu is set to **ExprOff** and **CoordOn**, then the Y= editor number of the function appears in the border at the top right of the screen, followed by the definition of the function.

If the Format menu is set to **ExprOff** and **CoordOff**, then tracing the graph is useless because all you see is a cursor moving on the graph. The calculator won't tell you the coordinates of the cursor location. (The Format menu and Y= editor are described in Chapter 9.)

If you've graphed more than one function and you want to trace a different function, press ⏶ or ⏷. Each time you press one of these keys, the cursor jumps to another function. Eventually it jumps back to the original function.

✔ **The values of *x* and *y*:** In the border at the bottom of the screen, you see the values of the *x*- and *y*-coordinates that define the cursor location. In the **PolarGC** format, the coordinates of this point display in polar form.

When you press TRACE, the cursor is placed on the graph at the point having an *x*-coordinate that is approximately midway between **Xmin** and **Xmax**. See the first screen in Figure 10-1. If the *y*-coordinate of the cursor location isn't between **Ymin** and **Ymax**, then the cursor doesn't appear on the screen. See the upcoming sidebar, "Panning in Function mode," to find out how to correct this situation.

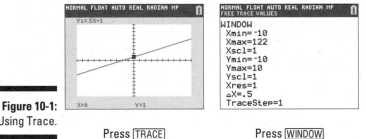

Figure 10-1: Using Trace.

Press TRACE   Press WINDOW

Each time you press ⏵, the cursor moves right to the next plotted point on the graph, and the coordinates of that point are displayed at the bottom of the screen. If you press ⏴, the cursor moves left to the previously plotted point. And if you press ⏶ or ⏷ to trace a different function, the tracing of that function starts at the point on the graph that has the *x*-coordinate displayed on-screen before you pressed this key.

Press CLEAR to terminate tracing the graph. This also removes the name of the function and the coordinates of the cursor from the screen.

Changing the TraceStep

The TI-84 Plus C allows you to change the *TraceStep*. The *TraceStep* is the amount the *x*-value changes each time you press ▶ or ◀ when tracing a function. The default TraceStep is approximately 0.1515151515. Customizing the TraceStep is easy and can be done by following these steps:

1. **Press WINDOW to access the Window editor.**

   See the second screen in Figure 10-1.

2. **Use ▼ to move your cursor the last line, titled TraceStep.**

3. **Enter your desired TraceStep.**

After pressing TRACE, your Trace cursor will move by the amount of the TraceStep value you entered each time you use ▶ or ◀.

Be careful, changing the TraceStep will automatically change the Xmax value as well. This can be quite a shock the first time it happens.

Moving the Trace cursor to any x-value in the graphing window

This is my favorite feature on the calculator! There's a hidden feature that works after you hit TRACE. See the first screen in Figure 10-2. If you want to start tracing your function at a specific value of the independent variable *x*, just key in that value and press ENTER when you're finished. (The value you assign to *x* must be between **Xmin** and **Xmax**; if it isn't, you get an error message.) When you're entering the *x*-value, your calculator displays the number you're entering in the border at the bottom of your screen, as shown in the second screen in Figure 10-2.

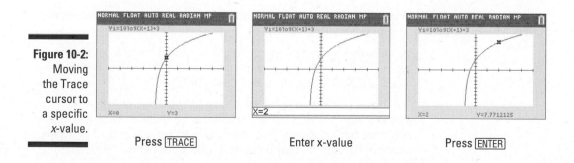

Figure 10-2:
Moving
the Trace
cursor to
a specific
x-value.

Press TRACE   Enter x-value   Press ENTER

Panning in Function mode

When you're tracing a function and the cursor hits the top or bottom of the screen, you will still see the coordinates of the cursor location displayed at the bottom of the screen but you won't see the cursor itself on the screen because the viewing window is too small. Press ENTER to get the calculator to adjust the viewing window to a viewing window that is centered about the cursor location. If the function you're tracing isn't displayed at the top of the screen, repeatedly press ▲ until it is. The Trace cursor then appears in the middle of the screen and you can use ▶ and ◀ to continue tracing the graph.

When you're tracing a function and the cursor hits the left or right side of the screen, the calculator automatically pans left or right. It also appropriately adjusts the values assigned to **Xmin** and **Xmax** in the Window editor — but it doesn't change the values of **Ymin** and **Ymax**, so you may not see the cursor on the screen. When this happens, press ENTER to make the calculator adjust the viewing window to one that's centered about the cursor location.

After you press ENTER, the Trace cursor moves to the point on the graph having the x-coordinate you just entered. See the third screen in Figure 10-2. This is an easy way to quickly substitute x-values into a function and see the output (y-values) as well as the nice visual of the Trace cursor on the graph itself. Pretty neat stuff, don't you think?

Using Zoom Commands

After you've graphed your functions (as described in Chapter 9), you can use Zoom commands to adjust the view of your graph. Press ZOOM to see the 17 Zoom commands that you can use. The following list explains the Zoom commands and how to use them:

✔ **Zoom commands that help you initially graph or regraph your function in a preset viewing window:**

• **ZStandard:** This command graphs your function in a preset viewing window where $-10 \leq x \leq 10$ and $-10 \leq y \leq 10$. You access it by pressing ZOOM 6. See the first screen in Figure 10-3.

This Zoom command is the best way to begin graphing. After graphing the function using **ZStandard**, you can, if necessary, use the **Zoom In** and **Zoom Out** commands to get a better idea of the nature of the graph. Using **Zoom In** and **Zoom Out** is described later in this section.

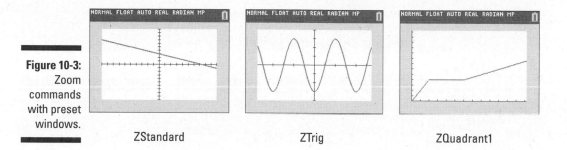

Figure 10-3:
Zoom
commands
with preset
windows.

ZStandard   ZTrig   ZQuadrant1

- **ZDecimal:** This command graphs your function in a preset viewing window where $-6.6 \leq x \leq 6.6$ and $-4.1 \leq y \leq 4.1$. The **ZDecimal** command is accessed by pressing [ZOOM][4].

  When you trace a function graphed in a **ZDecimal** window, the x-coordinate of the Trace cursor will be a multiple of 0.1.

- **ZTrig:** This command, which is most useful when graphing trigonometric functions, graphs your function in a preset viewing window where $-11\pi/4 \leq x \leq 11\pi/4$ and $-4 \leq y \leq 4$. It also sets the tick marks on the x-axis to multiples of $\pi/2$. You access **ZTrig** by pressing [ZOOM][7]. See the second screen in Figure 10-3.

  When you trace a function graphed in a **ZTrig** window, the x-coordinate of the Trace cursor will be a multiple of $\pi/24$.

- **ZQuadrant1:** This command graphs your function in a preset viewing window where $0 \leq x \leq 13.2$ and $0 \leq y \leq 13.2$. Of course, only Quadrant I can be viewed in this window. It is accessed by pressing [ZOOM][ALPHA][MATH], or by pressing [ZOOM] and using [▲] to scroll to **ZQuadrant1**. See the third screen in Figure 10-3.

  When you trace a function graphed in a **ZQuadrant1** window, the x-coordinate of the Trace cursor will be a multiple of 0.1.

To use the preceding zoom commands, enter your function into the calculator, press [ZOOM], and then press the key for the number of the command. The graph automatically appears.

✔ **Zoom commands that help you find an appropriate viewing window for the graph of your functions:**

- **ZoomFit:** This is a really neat Zoom command! If you know how you want to set the x-axis, **ZoomFit** automatically figures out the appropriate settings for the y-axis.

  To use **ZoomFit**, press [WINDOW] and enter the values you want for **Xmin**, **Xmax**, and **Xscl**. See the first screen in Figure 10-4. Then press [ZOOM][0] to get **ZoomFit** to figure out the y-settings and graph your function. See the second screen in Figure 10-4. **ZoomFit** does not figure out an appropriate setting for **Yscl**, so you may want to go back to the Window editor and adjust this value. The Window editor is discussed in Chapter 9.

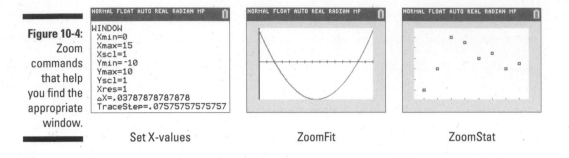

Figure 10-4: Zoom commands that help you find the appropriate window.

Set X-values   ZoomFit   ZoomStat

- **ZoomStat:** If you're graphing functions, this command is useless. But if you're graphing Stat Plots (as explained in Chapter 18), this command finds the appropriate viewing window for your plots. See the third screen in Figure 10-4.

✔ **Zoom commands that readjust the viewing window of an already graphed function:**

- **ZSquare:** Because the calculator screen isn't perfectly square, graphed circles won't look like real circles unless the viewing window is properly set. **ZSquare** readjusts the existing Window settings for you and then regraphs the function in a viewing window in which circles look like circles. Pictured in the first screen in Figure 10-5 is a circle in a ZStandard graphing window. See the difference a ZSquare window makes, as shown in the second screen in Figure 10-5.

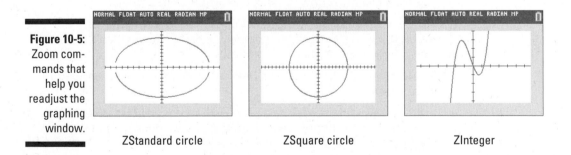

Figure 10-5: Zoom commands that help you readjust the graphing window.

ZStandard circle   ZSquare circle   ZInteger

To use **ZSquare**, graph the function as described in Chapter 9, and then press [ZOOM][5]. The graph automatically appears.

- **ZInteger:** This command is quite useful when you want the Trace cursor to trace your functions using integer values of the *x*-coordinate, such as when graphing a function that defines a sequence. **ZInteger** readjusts the existing Window settings and regraphs the function in a viewing window in which the Trace cursor displays integer values for the *x*-coordinate. In addition, **ZInteger** sets the **Xscl** and **Yscl** to 10.

  To use **ZInteger**, graph the function as described in Chapter 9, and then press ZOOM 8. Use the ▶ ◀ ▲ ▼ keys to move the cursor to the spot on the screen that will become the center of the new screen. Then press ENTER. The graph is redrawn centered at the cursor location. See the third screen in Figure 10-5.

✔ **Zoom commands that zoom in or zoom out from an already graphed function:**

- **Zoom In and Zoom Out:** After the graph is drawn (as described in Chapter 9), these commands enable you to zoom in on a portion of the graph or to zoom out from the graph. They work very much like a zoom lens on a digital camera.

  Press ZOOM 2 to zoom in or press ZOOM 3 to zoom out. Then use the ▶ ◀ ▲ ▼ keys to move the cursor (the cursor looks like a + sign) to the spot on the screen from which you want to zoom in or zoom out. Then press ENTER. The graph is redrawn centered at the cursor location.

  You can press ENTER again to zoom in closer or to zoom out one more time. Press CLEAR when you're finished zooming in or zooming out. You may have to adjust the window settings, as described in Chapter 9.

- **ZBox:** Some functions end up having really interesting graphs. One such function is shown in the Y= editor in the first screen in Figure 10-6. The ZBox command enables you to define a new viewing window for a portion of your graph by enclosing it in a box, as illustrated in the second screen in Figure 10-6. Looking at the second screen in Figure 10-6, the function appears to be a normal cosine wave. The box becomes the new viewing window as shown in the third screen in Figure 10-6. After using ZBox to take a closer look at the function, it's easy to see that this isn't an ordinary cosine wave!

  To construct the box, press ZOOM 1 and use the ▶ ◀ ▲ ▼ keys to move the cursor (the cursor looks like a + sign) to the spot where you want one corner of the box to be located. Press ENTER to anchor that corner of the box. Then use the ▶ ◀ ▲ ▼ keys to construct the rest of the box. When you press these keys, the calculator draws the sides of the box. Press ENTER when you're finished drawing the box. The graph is then redrawn in the window defined by your box.

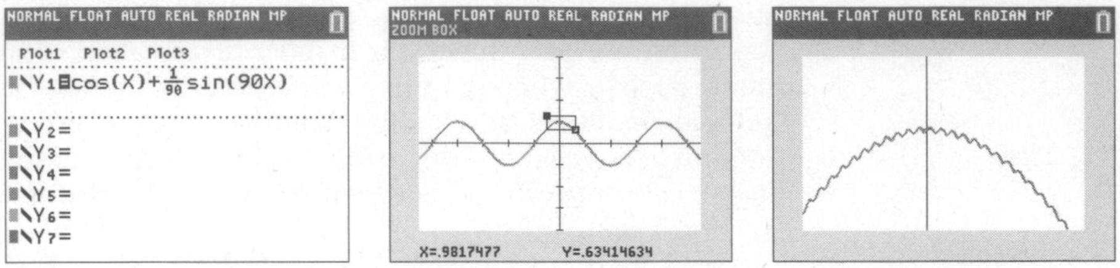

Figure 10-6:
Zoom commands that help you zoom in or out.

| Interesting function | Draw the box | ZBox result |

When you use **ZBox**, if you don't like the size of the box you get, you can use any of the ▶◀▲▼ keys to resize the box. If you don't like the location of the corner you anchored, press CLEAR and start over.

When you use **ZBox**, ENTER is pressed only two times. The first time you press it is to anchor a corner of the zoom box. The next time you press ENTER is when you're finished drawing the box, and you're ready to have the calculator redraw the graph.

✔ **Zoom commands that enable you to trace by fraction steps:**

- **ZFrac1/2:** This command graphs your function in a preset viewing window where $-66/2 \leq x \leq 66/2$ and $-41/2 \leq y \leq 41/2$. It is accessed by pressing ZOOM ALPHA APPS, or by pressing ZOOM and using ▲ to scroll to **ZFrac1/2**.

  When you trace a function graphed in a **ZFrac1/2** window, the x-coordinate of the Trace cursor will be a multiple of 1/2. I love to see improper fractions as x-coordinates. See "Tracing a Graph" section which appears earlier in this chapter for more details.

- **ZFrac1/3, ZFrac 1/4, ZFrac1/5, ZFrac 1/8, ZFrac1/10:** These commands graph your function in a preset viewing window and work in the same manner as **ZFrac1/2** does. If you think of d as the denominator of your fraction, then the viewing window is $-66/d \leq x \leq 66/d$ and $-41/d \leq y \leq 41/d$. Tracing with these commands enables you to trace the x-coordinates by multiples of $1/d$.

Undoing a zoom

If you use a Zoom command to redraw a graph and then want to undo what that command did to the graph, follow these steps:

1. Press ZOOM ▶ **to access the Zoom Memory menu.**

2. Press 1 **to select ZPrevious.**

The graph is redrawn as it appeared in the previous viewing window.

Storing and recalling your favorite graphing window

You can make a preset graphing window of your own! Maybe you (or your teacher) have a favorite setting for a graphing window? For example, I performed a ZOOM 6 followed by a ZOOM 5 to get the window that appears in the first screen in Figure 10-7. I like this window because it doesn't distort circles and it's large enough to see most functions. Follow the steps below to store and recall any graphing window that you happen to like:

1. **Press** ZOOM ▶ **to access the Zoom MEMORY menu.**

   See the second screen in Figure 10-7.

Figure 10-7:
Storing and recalling your favorite graphing window.

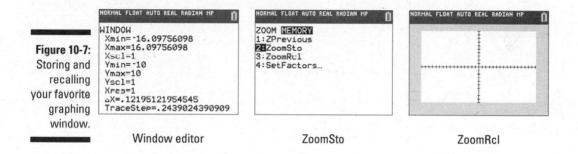

Window editor   ZoomSto   ZoomRcl

2. **Press** 2 **to store your graphing window.**

   Most of the variables in the Window editor are saved; the exceptions are the **ΔX** and **TraceStep** values (which are not stored). Even if you turn your calculator off, your graphing window will remain stored in Zoom MEMORY.

3. **Press** ZOOM ▶ 3 **to recall your graphing window.**

   See the third screen in Figure 10-7.

Displaying Functions in a Table

After you've entered the functions in the Y= editor, you can have the calculator create a table of functional values. I love that the table values are automatically color-coded to match the color of the functions on the graph. There are two kinds of tables you can create: an automatically generated table and a user-generated table.

Automatically generated table

To automatically generate a table, perform the following steps:

1. **Highlight the equal sign of those functions in the Y= editor that you want to appear in the table.**

   Only those functions in the Y= editor that are defined with a highlighted equal sign appear in the table. To highlight or remove the highlight from an equal sign, press Y=, use the ▶◀▲▼ keys to place the cursor on the equal sign in the definition of the function, and then press ENTER to toggle the equal sign between highlighted and not highlighted. See the first screen in Figure 10-8.

Figure 10-8:
Automatically generating a table.

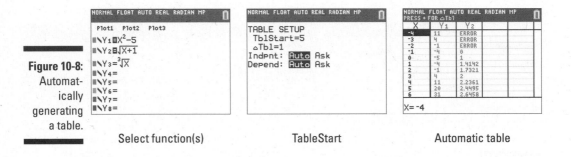

Select function(s)   TableStart   Automatic table

2. **Press 2nd WINDOW to access the Table Setup editor.**

3. **Enter a number in TblStart, and then press ENTER.**

   TblStart is the first value of the independent variable *x* to appear in the table. In the second screen in Figure 10-8, **TblStart** is assigned the value 5.

   To enter the number you have chosen for **TblStart**, place the cursor on the number appearing after the equal sign, press the number keys to enter your new number, and then press ENTER.

4. **Enter a number in ΔTbl, and then press ENTER.**

   ΔTbl gives the increment for the independent variable *x*. In the second screen in Figure 10-8, **ΔTbl** is assigned the value –1.

   To enter the number you have chosen for **ΔTbl**, place the cursor on the number appearing after the equal sign, press the number keys to enter your new number, and then press ENTER.

5. **Press 2nd GRAPH to display the table.**

   See the third screen in Figure 10-8. Here's what you see and what you can do with an automatically generated table:

   • If **Indpnt** and **Depend** are both in **Auto** mode, then when you press 2nd GRAPH, the table is automatically generated. To display rows in the table beyond the last row on the screen, repeatedly press ⊡ until they appear. You can repeatedly press ⊡ to display rows above the first row on the screen.

- Notice the Context Help message "Press + for ΔTbl" in the border at the top of the third screen in Figure 10-8. If the TableStep is not to your liking, press ⊞, enter your new TableStep, and press ENTER.

- If you're constructing a table for more than four functions, only the first four functions appear on the screen. To see the other functions, repeatedly press ▶ until they appear. This causes one or more of the initial functions to disappear. To see them again, repeatedly press ◀ until they appear.

- Each time the calculator redisplays a table with a different set of rows, it also automatically resets **TblStart** to the value of *x* that appears in the first row of the newly displayed table. To return the table to its original state, press 2nd WINDOW to access the Table Setup editor, and then change the value that the calculator assigned to **TblStart**.

User-generated table

To create a user-generated table, perform the following steps:

1. Press 2nd WINDOW to access the Table Setup editor.

2. Set the mode for Indpnt and Depend.

To change the mode of either **Indpnt** or **Depend**, use the ▶◀▲▼ keys to place the cursor on the desired mode, either **Auto** or **Ask**, and then press ENTER.

I recommend putting **Indpnt** in **Ask** mode and **Depend** in **Auto** mode, as shown in the first screen in Figure 10-9.

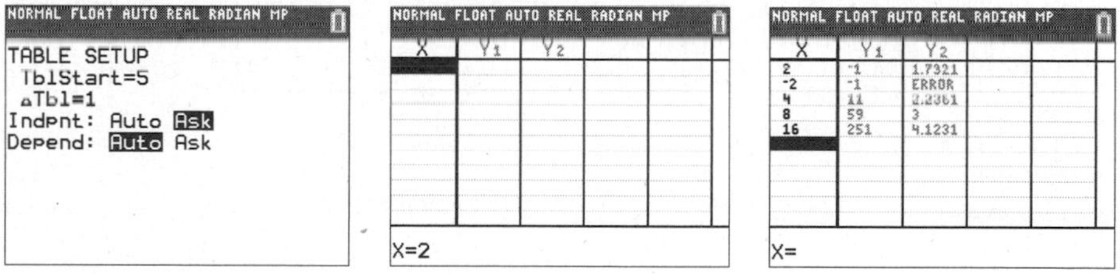

Figure 10-9:
A user-generated table.

Table Setup editor	Enter Indpnt value	Error in table

3. Press 2nd GRAPH to display the table.

When you display the table, it should be empty. If it's not empty, clear the table (see the "Clearing a Table" section later in this chapter).

In an empty table, key in the first value of the independent variable x that you want to appear in the table, as shown in the second screen in Figure 10-9. Press ENTER and the corresponding y-values of the functions in the table automatically appear. Key in the next value of x you want in the table and press ENTER, and so on. The values of x that you place in the first column of the table don't have to be in any specific order, nor do they have to be between the **Xmin** and **Xmax** settings in the Window editor!

For a user-defined table, you don't have to assign values to **TblStart** and **ΔTbl** in the Table Setup editor.

The other combinations of mode settings for **Indpnt** and **Depend** are not all that useful, unless you want to play a quick round of "Guess the y-coordinate."

The word ERROR appearing in a table doesn't indicate that the creator of the table has done something wrong. It indicates that either the function is undefined or the corresponding value of x is not a real-valued number. This is illustrated in the third screen in Figure 10-9.

Editing a function in a table

While displaying the table of functional values, you can edit the definition of a function without going back to the Y= editor. To do this, use the ▶◀▲▼ keys to place the cursor on the column heading for that function and then press ENTER. See the first screen in Figure 10-10.

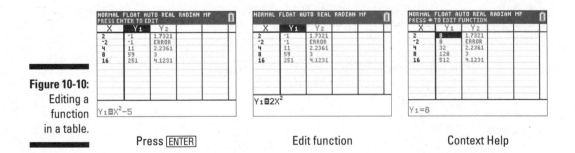

Figure 10-10: Editing a function in a table.

Press ENTER   Edit function   Context Help

Edit the definition of the function as shown in the second screen in Figure 10-10. Press ENTER when you're finished. The calculator automatically updates the table and the definition of the function in the Y= editor.

Context Help in the Status bar in the border at the top of the screen gives helpful reminders. See the Context Help reminder, "Press ⬆ to edit function," in the third screen in Figure 10-10.

Clearing a Table

Not all tables are created alike. An automatically generated table, for example, cannot be cleared. To change the contents of such a table, you have to change the values assigned to **TblStart** and **ΔTbl** in the Table Setup editor. After you have created a user-defined table, however, you can perform the following four steps to clear its contents:

1. **Press 2nd WINDOW to access the Table Setup editor and then set Indpnt to Auto.**

2. **Press 2nd GRAPH to display an automatically generated table.**

3. **Press 2nd WINDOW and set Indpnt back to Ask.**

4. **Press 2nd GRAPH to display an empty table.**

If these steps seem a little repetitive to you, there's another way to clear a table. Just follow these steps:

1. **Press 2nd MODE to access the Home screen.**

2. **Press 2nd 0 to access the Catalog.**

3. **Press PRGM to jump to the commands starting with the letter C, then use ⬇ to scroll to the ClrTable.**

See the first screen in Figure 10-11.

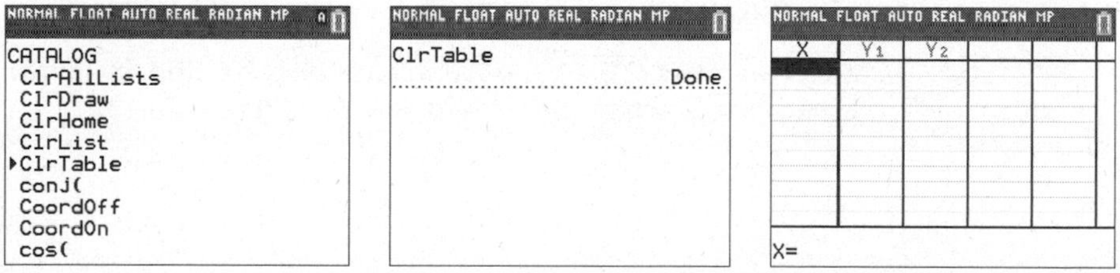

Figure 10-11:
Clearing a table.

Catalog   Press ENTER   Cleared table

4. **Press** ENTER **to insert the ClrTable command, and then press** ENTER **again to clear the table.**

   See the second screen in Figure 10-11.

5. **Press** 2nd GRAPH **to display the newly cleared table.**

   See the third screen in Figure 10-11.

Viewing the Table and the Graph on the Same Screen

After you have graphed your functions and created a table of functional values, you can view the graph and the table on the same screen. To do so, follow these steps:

1. **Press** MODE.

2. **Put the calculator in Graph-Table mode.**

   To do so, use the ▶ ◀ ▲ ▼ keys to place the cursor on **Graph-Table** in the ninth line of the Mode menu, and then press ENTER to highlight it. This is illustrated in the first screen in Figure 10-12.

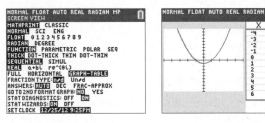

Figure 10-12: A Graph-Table split screen.

Graph-Table mode   Press GRAPH

3. **Press** GRAPH.

   After you press GRAPH, the graph and the table appear on the same screen (as shown in the second screen in Figure 10-12).

If you press any key used in graphing functions, such as GRAPH or TRACE, the cursor becomes active on the graph side of the screen. To return the cursor to the table, press 2nd GRAPH. See the first screen in Figure 10-13.

In Graph-Table mode, only one function will display in the table at a time. If you have more than one function graphed, press the ▶ key to see additional table values. For this to work, your cursor must be on the table side of the screen.

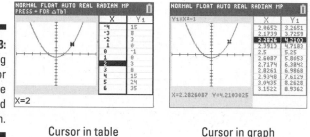

Figure 10-13:
Moving the cursor between the table and graph.

Cursor in table   Cursor in graph

If you press TRACE and then use the ▶◀▲▼ keys to trace the graph, the value of the independent variable *x* corresponding to the cursor location on the graph is highlighted in the table and the column for the function you're tracing appears next to it. If necessary, the calculator updates the table so you can see that row in the table.

Press WINDOW to change your TraceStep settings. When your cursor is on the graph side of the screen, your table values are determined by your TraceStep, as shown in the second screen in Figure 10-13.

To view the graph or the table in full screen mode, you can use these steps:

1. **Press** MODE.

2. **Put the calculator in Full screen mode.**

   To do so, use the ▶◀▲▼ keys to place the cursor on **Full** in the bottom-left corner of the Mode menu and press ENTER to highlight it.

3. **Press** GRAPH **to see the graph, or press** 2nd GRAPH **to see the table.**

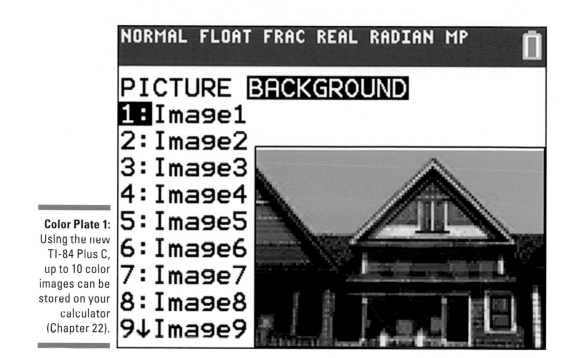

Color Plate 1: Using the new TI-84 Plus C, up to 10 color images can be stored on your calculator (Chapter 22).

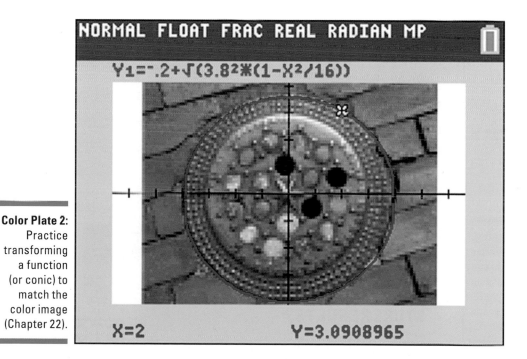

Color Plate 2: Practice transforming a function (or conic) to match the color image (Chapter 22).

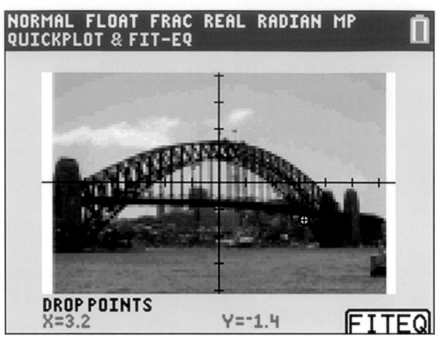

Color Plate 3: Use Quick Plot to place points directly on a graph (Chapter 22).

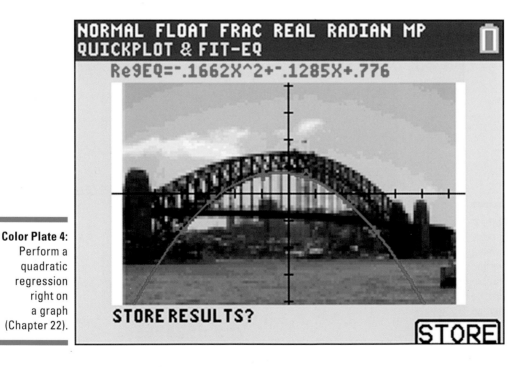

Color Plate 4: Perform a quadratic regression right on a graph (Chapter 22).

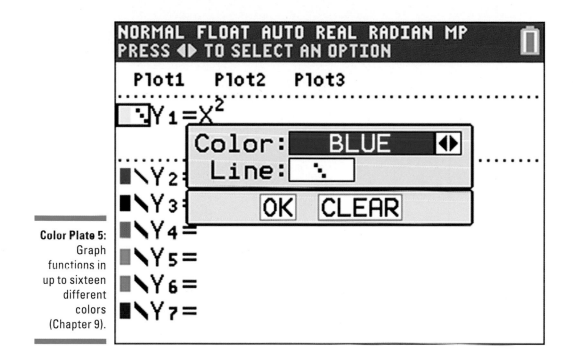

Color Plate 5: Graph functions in up to sixteen different colors (Chapter 9).

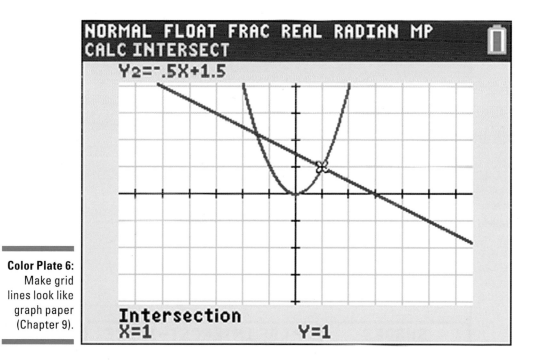

Color Plate 6: Make grid lines look like graph paper (Chapter 9).

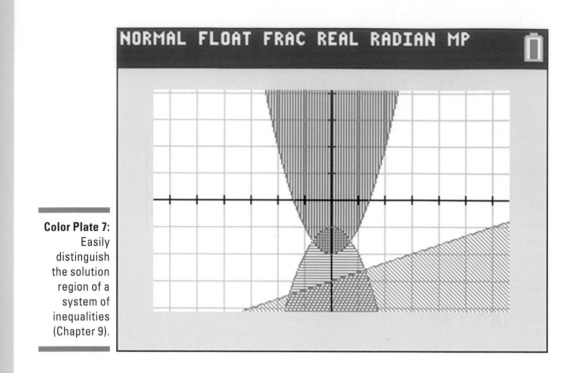

Color Plate 7:
Easily distinguish the solution region of a system of inequalities (Chapter 9).

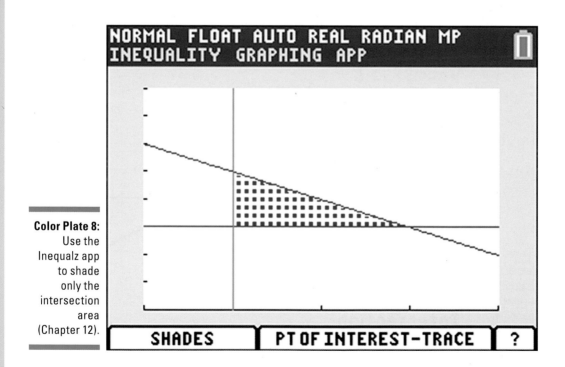

Color Plate 8:
Use the Inequalz app to shade only the intersection area (Chapter 12).

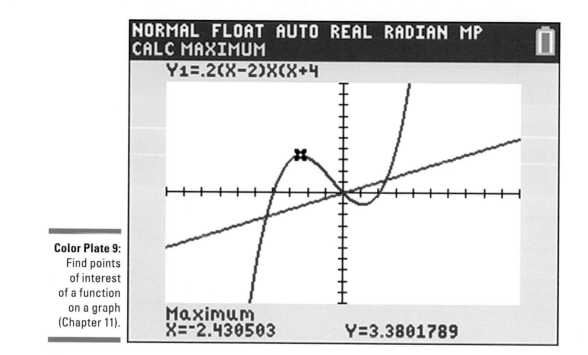

Color Plate 9:
Find points
of interest
of a function
on a graph
(Chapter 11).

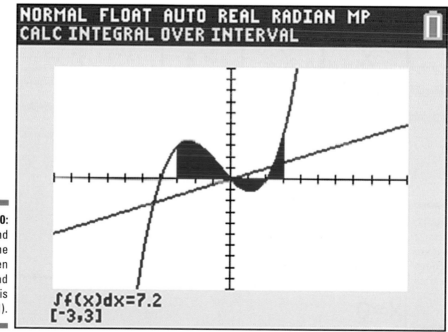

Color Plate 10:
Calculate and
display the
area between
a curve and
the x-axis
(Chapter 11).

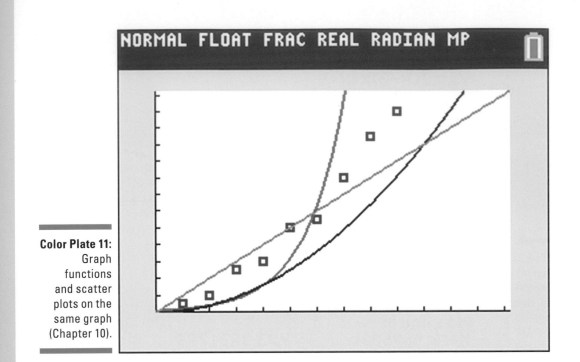

Color Plate 11: Graph functions and scatter plots on the same graph (Chapter 10).

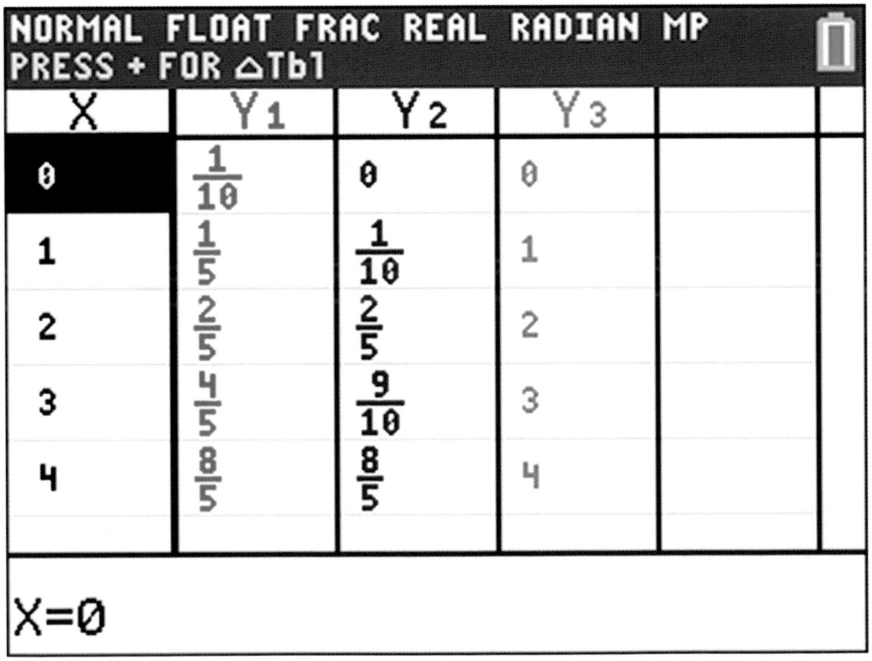

X	Y₁	Y₂	Y₃		
0	$\frac{1}{10}$	0	0		
1	$\frac{1}{5}$	$\frac{1}{10}$	1		
2	$\frac{2}{5}$	$\frac{2}{5}$	2		
3	$\frac{4}{5}$	$\frac{9}{10}$	3		
4	$\frac{8}{5}$	$\frac{8}{5}$	4		

X=0

Color Plate 12: Customize lists that are color-coded to match the function (Chapter 10).

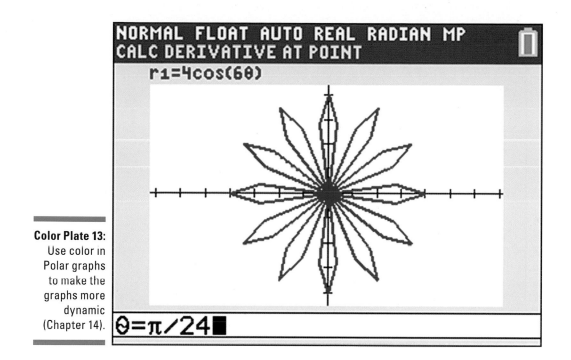

Color Plate 13: Use color in Polar graphs to make the graphs more dynamic (Chapter 14).

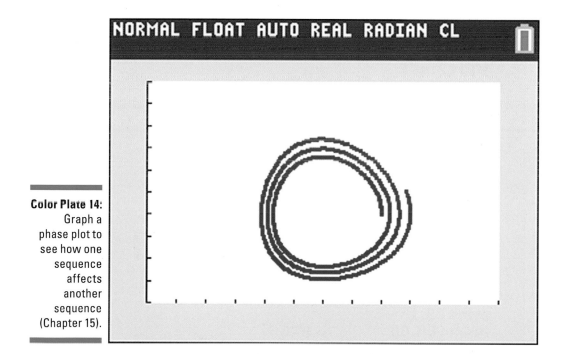

Color Plate 14: Graph a phase plot to see how one sequence affects another sequence (Chapter 15).

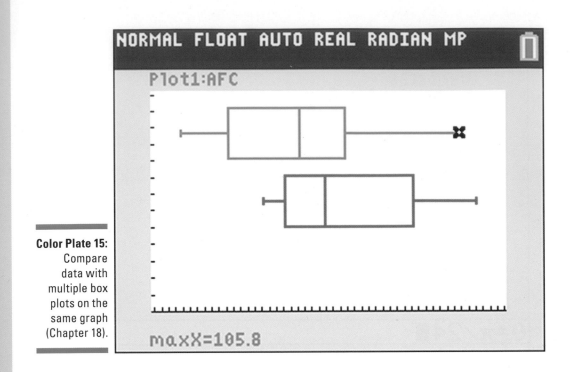

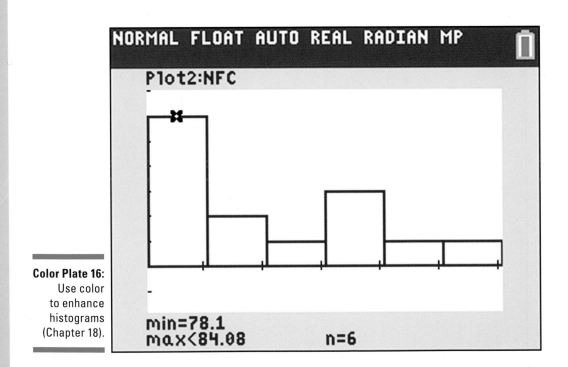

Chapter 11

Evaluating Functions

A fter graphing a function (as described in Chapter 9), you can use the options on the Calculate menu to find the value of the function at a specified value of x, to find the zeros (x-intercepts) of the function, and to find the maximum and minimum values of the function. You can even find the derivative of the function at a specified value of x, or you can evaluate a definite integral of the function. This, in turn, enables you to find the slope of the tangent to the graph of the function at a specified value of x or to find the area between the graph and the x-axis. Moreover, if you have graphed two functions, there's an option on the Calculate menu that finds the coordinates of these two functions' points of intersection.

The rest of this chapter tells you how to use the Calculate menu to find these values. But be warned: The calculator is not perfect. In most cases, using the options on the Calculate menu yields only an approximation of the true value (albeit a very *good* approximation). Before using the Calculate menu, double-check that your Format menu ([2nd][ZOOM]) is set to ExprOn and CoordOn.

Finding the Value of a Function

If you want to substitute a value in a function, you could accomplish this task by using paper and a pencil. However, wouldn't it be easier to use your calculator to find the value of a function? There are a few different ways to accomplish this task.

The TI-84 Plus C displays functions and information in the border of the graph screen. The TI-84 Plus displays similar information directly on the graph screen.

Using your graph to find the value of a function

The **CALC** menu can be used to evaluate a function at any specified *x*-value. To access and use this command, perform the following steps:

1. **Graph the functions in a viewing window that contains the specified value of *x*.**

   Graphing functions and setting the viewing window are explained in Chapter 9. To get a viewing window containing the specified value of *x*, that value must be between **Xmin** and **Xmax**.

2. **Press** 2nd TRACE **to access the Calculate menu.**

3. **Press** ENTER **to select the value option.**

4. **Enter the specified value of *x*.**

   When using the **value** command to evaluate a function at a specified value of *x*, that value must be an *x*-value that appears on the *x*-axis of the displayed graph — that is, it must be between **Xmin** and **Xmax**. If it isn't, you get an error message.

   Use the keypad to enter the value of *x* (as illustrated in the first screen in Figure 11-1). If you make a mistake when entering your number, press CLEAR and re-enter the number.

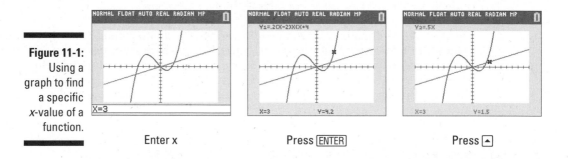

Figure 11-1:
Using a graph to find a specific *x*-value of a function.

Enter x   Press ENTER   Press ▲

5. **Press** ENTER.

   After you press ENTER, the first highlighted function in the Y= editor appears in the border at the top of the screen, the cursor appears on the graph of that function at the specified value of *x*, and the coordinates of the cursor appear in the border at the bottom of the screen. See the second screen in Figure 11-1.

You can also find the value of a function by pressing [TRACE], entering an *x*-value, and pressing [ENTER].

6. **Repeatedly press the [▲][▼] keys to see the value of the other graphed functions at your specified value of *x*.**

   Each time you press the [▲][▼]keys, the name of the function being evaluated appears in the border at the top of the screen and the coordinates of the cursor location appears in the border at the bottom of the screen. This is illustrated in the third screen in Figure 11-1.

After using the **value** command to evaluate your functions at one value of *x*, you can evaluate your functions at another value of *x* by keying in the new value and pressing [ENTER]. Pressing any function key (such as [ENTER] or [TRACE]) *after* evaluating a function deactivates the **value** command.

If you plan to evaluate functions at several specified values of *x*, consider constructing a user-defined table of functional values (as explained in Chapter 10).

Using your calculator to find the value of a function

Another way to find the value of a function involves using your calculator. This method is easy and doesn't have the restrictions the graphing method has (the *x*-value has to be between the Xmin and Xmax).

Follow these steps to use your calculator to find the value of a function:

1. **Enter your function in the Y= editor.**

   You need to remember the name of the function you enter. I entered an equation in Y_1 as shown in the first screen in Figure 11-2.

Figure 11-2:
Using your calculator to find a specific *x*-value of a function.

```
NORMAL FLOAT AUTO REAL RADIAN MP
Plot1  Plot2  Plot3
■\Y1■.2X(X-2)(X+4)
■\Y2=
■\Y3=
■\Y4=
■\Y5=
■\Y6=
■\Y7=
■\Y8=
■\Y9■
```
Y= editor

```
NORMAL FLOAT AUTO REAL RADIAN MP

                    1:Y1 6:Y6
                    2:Y2 7:Y7
                    3:Y3 8:Y8
                    4:Y4 9:Y9
                    5:Y5 0:Y0
FRAC FUNC MTRX YVAR
```
Y-VAR menu

```
NORMAL FLOAT AUTO REAL RADIAN MP
Y1(3)
                              4.2
```
Press [ENTER]

2. **Press** 2nd MODE **to access the Home screen.**

3. **Press** ALPHA TRACE **to access the Y-VAR menu and choose the function you need.**

   See the second screen in Figure 11-2.

4. **Press** (**and enter the *x*-value you would like evaluated.**

5. **Press**) **and then press** ENTER.

   See the third screen in Figure 11-2.

Composing Functions

Sometimes functions are composed together. In your textbook, this may look like, $f(g(x))$. Function composition is really just substituting one function into another function. Fortunately, you can use your calculator to accomplish this task.

Using your graph to compose functions

If you want a graphical representation of function composition, follow these steps:

1. **Enter your functions in the Y= editor.**

   I entered my functions in Y_1 and Y_2 as shown in the first screen in Figure 11-3.

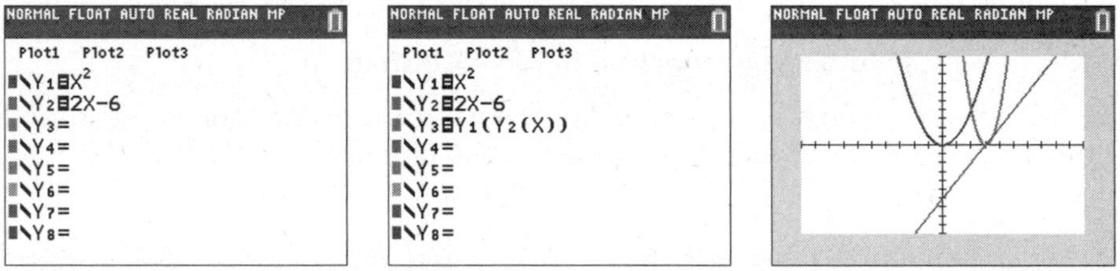

Figure 11-3: Graphing composed functions.

Y= editor

Enter function

Press GRAPH

2. **Use** ▶ ◀ ▲ ▼ **to place your cursor in an open equation in the Y= editor.**

3. **Press** ALPHA TRACE **to access the Y-VAR menu and choose the first function you need.**

4. **Press** (**and press** ALPHA TRACE **to access the Y-VAR menu and choose the second function you need.**

5. **Press** (**, then press** X,T,Θ,n **and press**) **twice.**

   See the second screen in Figure 11-3.

6. **Press** GRAPH **to see the graph of the composed function.**

   See the third screen in Figure 11-3.

Using your calculator to compose functions

To evaluate composed functions at a specific x-value, follow these steps:

1. **Enter your functions in the Y= editor.**

   I entered my functions in Y_1 and Y_2 as shown in the first screen in Figure 11-4.

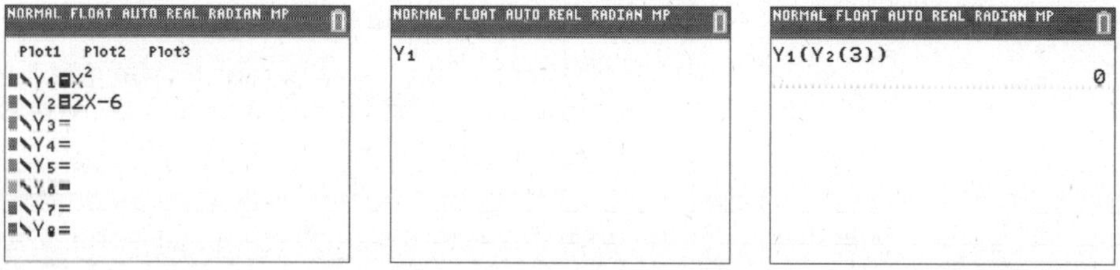

Figure 11-4:
Function
composition
at a specific
x-value.

NORMAL FLOAT AUTO REAL RADIAN MP	NORMAL FLOAT AUTO REAL RADIAN MP	NORMAL FLOAT AUTO REAL RADIAN MP
Plot1 Plot2 Plot3 ∎\Y₁▪X² ∎\Y₂▪2X-6 ∎\Y₃= ∎\Y₄= ∎\Y₅= ∎\Y₆▪ ∎\Y₇= ∎\Y₈=	Y₁	Y₁(Y₂(3)) 0
Y= editor	Enter function	Press ENTER

2. **Press** 2nd MODE **to access the Home screen.**

3. **Press** ALPHA TRACE **to access the Y-VAR menu and choose the first function you need.**

   See the second screen in Figure 11-4.

4. **Press** (**and press** ALPHA TRACE **to access the Y-VAR menu and choose the second function you need.**

5. **Press** (**, then enter an** x**-value and press**) **twice.**

6. **Press** ENTER **to see the result of your function composition.**

   See the third screen in Figure 11-4.

Finding the Zeros of a Function

The *zeros* of the function $y = f(x)$ are the solutions to the equation $f(x) = 0$. Because $y = 0$ at these solutions, these zeros (solutions) are really just the *x*-coordinates of the *x*-intercepts of the graph of $y = f(x)$. (An *x*-intercept is a point where the graph crosses or touches the *x*-axis.)

To find a zero of a function, perform the following steps:

1. **Graph the function in a viewing window that contains the zeros of the function.**

   Graphing a function and finding an appropriate viewing window are explained in Chapter 9. To get a viewing window containing a zero of the function, that zero must be between **Xmin** and **Xmax** and the *x*-intercept at that zero must be visible on the graph.

2. **Press** [2nd][TRACE] **to access the Calculate menu.**

3. **Press** [2] **to select the zero option.**

4. **If necessary, repeatedly press the** [▲][▼] **keys until the appropriate function appears in the border at the top of the screen.**

5. **Set the Left Bound for the zero you desire to find.**

   To do so, use the [◄][►] keys to place the cursor on the graph a little to the left of the zero, and then press [ENTER]. Alternatively, you can enter a number and press [ENTER] to establish the Left Bound.

   On the TI-84 Plus C, a Left Bound vertical line appears on the screen (as illustrated by the dotted line with a small triangular indicator in the first screen of Figure 11-5).

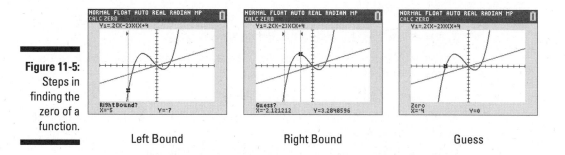

Figure 11-5: Steps in finding the zero of a function.

Left Bound   Right Bound   Guess

6. **Set the Right Bound for the zero.**

To do so, use the ⊲▸ keys to place the cursor on the graph a little to the right of the zero, and then press ENTER. Alternatively, you can enter a number and press ENTER to establish the Right Bound.

On the TI-84 Plus C, a Right Bound dotted line with a small triangular indicator appears on the screen, as shown in the second screen of Figure 11-5.

7. **Tell the calculator where you guess the zero is located.**

This guess is necessary because the calculator uses a numerical routine for finding a zero. The routine is an iterative process that requires a seed (guess) to get it started. The closer the seed is to the zero, the faster the routine finds the zero. To do this, use the ⊲▸ keys to place the cursor on the graph as close to the zero as possible, and then press ENTER. The value of the zero appears in the border at the bottom of the screen, as shown in the third screen of Figure 11-5.

The calculator uses scientific notation to denote really large or small numbers. For example, -0.00000001 is displayed on the calculator as $-1E-8$, and 0.000000005 is displayed as $5E-8$.

Finding Min and Max

Finding the maximum or minimum point on a graph has many useful applications. For example, the maximum point on the graph of a profit function tells you not only the maximum profit (the y-coordinate), but also how many items (the x-coordinate) the company must manufacture to achieve this profit. To find the minimum or maximum value of a function, perform the following steps:

1. **Graph the function in a viewing window that contains the minimum and/or maximum values of the function.**

Graphing a function and finding an appropriate viewing window are explained in Chapter 9.

2. **Press 2nd TRACE to access the Calculate menu.**

3. **Press 3 to find the minimum, or press 4 to find the maximum.**

4. **If necessary, repeatedly press the ▲▼ keys until the appropriate function appears in the border at the top of the screen.**

5. **Set the Left Bound of the minimum or maximum point.**

To do so, use the ⊲▸ keys to place the cursor on the graph a little to the left of the location of the minimum or maximum point, and then press ENTER. A *Left Bound indicator* (the dotted line with a triangular indicator shown in the first screen of Figure 11-6) appears on the screen.

Figure 11-6:
Steps in finding the maximum value of a function.

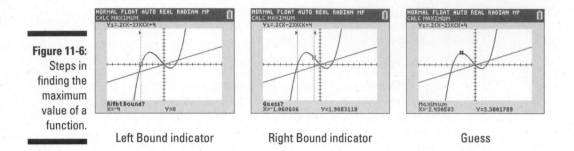

Left Bound indicator   Right Bound indicator   Guess

6. **Set the Right Bound for the zero.**

   To do so, use the ◄|► keys to place the cursor on the graph a little to the right of the location of the minimum or maximum point, and then press ENTER. A *Right Bound indicator* (the rightmost dotted line with the triangular indicator in the second screen of Figure 11-6) appears on the screen.

7. **Tell the calculator where you guess the min or max is located.**

   To do so, use the ◄|► keys to place the cursor on the graph as close to the location of the minimum or maximum point as possible, and then press ENTER. The coordinates of the minimum or maximum point appears in the border at the bottom of the screen (as shown in the third screen of Figure 11-6).

Finding Points of Intersection

Using the ►|◄|▲|▼ keys in a graph activates a free-moving trace. However, using a free-moving trace rarely locates the point of intersection of two graphs but instead gives you an *approximation* of that point. To accurately find the coordinates of the point where two functions intersect, perform the following steps:

1. **Graph the functions in a viewing window that contains the point of intersection of the functions.**

   Graphing a function and finding an appropriate viewing window are explained in Chapter 9.

2. **Press 2nd TRACE to access the Calculate menu.**

3. **Press 5 to select the intersect option.**

4. **Select the first function.**

   If the name of one of the intersecting functions does not appear in the border at the top of the screen, repeatedly press the ▲▼ keys until it does. This is illustrated in the first screen in Figure 11-7. When the cursor is on one of the intersecting functions, press ENTER to select it.

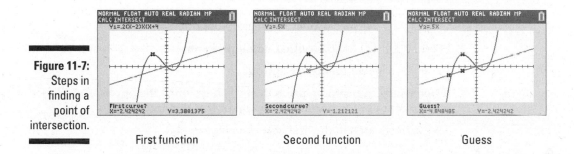

Figure 11-7:
Steps in
finding a
point of
intersection.

First function   Second function   Guess

5. **Select the second function.**

   If the calculator does not automatically display the name of the second intersecting function in the border at the top of the screen, repeatedly press the ▲▼ keys until it does. This is illustrated in the second screen in Figure 11-7. When the cursor is on the second intersecting function, press ENTER to select it.

6. **Use the ◄ ► keys to move the cursor as close to the point of intersection as possible.**

   This is illustrated in the third screen in Figure 11-7.

7. **Press ENTER to display the coordinates of the point of intersection.**

If there are only two functions in the Y= editor, you can save time by pressing 2nd TRACE ENTER ENTER to choose your functions. If there is only one point of intersection of the two functions, then press ENTER again to calculate the point of intersection. It is only necessary to make a guess when there is more than one point of intersection.

Finding the Slope of a Curve

The calculator is not equipped to find the derivative of a function. For example, it can't tell you that the derivative of x^2 is $2x$. But the calculator is equipped with a numerical routine that evaluates the derivative at a specified value of x.

This numerical value of the derivative is the slope of the tangent to the graph of the function at the specified x-value. It is also called the slope of the curve. To find the slope (derivative) of a function at a specified value of x, perform the following steps:

1. **Graph the function in a viewing window that contains the specified value of x.**

   Graphing a function and setting the viewing window are explained in Chapter 9. To get a viewing window containing the specified value of x, that value must be between **Xmin** and **Xmax**.

2. **Press [2nd][TRACE] to access the Calculate menu.**

3. **Press [6] to select the dy/dx option.**

4. **If necessary, repeatedly press the [▲][▼] keys until the appropriate function appears in the border at the top of the screen.**

   This is illustrated in the first screen in Figure 11-8.

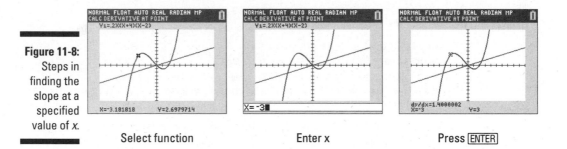

Figure 11-8: Steps in finding the slope at a specified value of x.

Select function   Enter x   Press [ENTER]

5. **Enter the specified value of x.**

   To do so, use the keypad to enter the value of x. As you use the keypad, **X=** appears, replacing the coordinates of the cursor location appearing at the bottom of the screen in Step 4. The number you key in appears after **X=**. This is illustrated in the second screen in Figure 11-8. If you make a mistake when entering your number, press [CLEAR] and re-enter the number.

   If you are interested only in finding the slope of the function in a general area of the function instead of at a specific value of x, instead of entering a value of x, just use the [◄] and [►] to move the cursor to the desired location on the graph of the function.

6. **Press [ENTER].**

   After pressing [ENTER], the slope (derivative) is displayed in the border at the bottom of the screen. This is illustrated in the third screen in Figure 11-8.

Evaluating a Definite Integral

If $f(x)$ is positive for $a \leq x \leq b$, and then the definite integral $\int_a^b f(x)\,dx$ also gives the area between the curve and the x-axis for $a \leq x \leq b$. To evaluate the definite integral, perform the following steps:

1. **Graph the function $f(x)$ in a viewing window that contains the Lower Limit a and the Upper Limit b.**

   Graphing a function and setting the viewing window are explained in Chapter 9. To get a viewing window containing a and b, these values must be between **Xmin** and **Xmax**.

2. **Press** 2nd TRACE **to access the Calculate menu.**

3. **Press** 7 **to select the** $\int f(x)\,dx$ **option.**

4. **If necessary, repeatedly press the** ▲▼ **keys until the appropriate function appears in the border at the top of the screen.**

   This process is illustrated in the first screen in Figure 11-9.

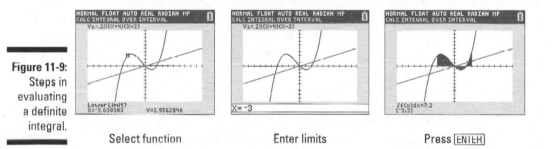

Figure 11-9:
Steps in
evaluating
a definite
integral.

Select function   Enter limits   Press ENTER

5. **Enter the value of the Lower Limit a.**

   To do so, use the keypad to enter the value of the Lower Limit a. As you use the keypad, **X=** appears, replacing the coordinates of the cursor location appearing at the bottom of the screen in Step 4. The number you key in appears after **X=**. This is illustrated in the second screen in Figure 11-9. If you make a mistake when entering your number, press CLEAR and re-enter the number.

6. **Press** ENTER.

   After pressing ENTER, a *Left Bound indicator* (the dotted line with a triangular indicator) appears on the graphing screen.

7. **Enter the value of the Upper Limit b and press** ENTER.

After pressing ENTER, the value of the definite integral appears in the border at the bottom of the screen and the area between the curve and the *x*-axis, for $a \leq x \leq b$, will be shaded. This is illustrated in the third screen in Figure 11-9. I really like how the TI-84 Plus C uses interval notation to display the interval of the definite integral.

The shading of the graph produced by using the $\int f(x)\ dx$ option on the Calculate menu doesn't automatically vanish when you use another Calculate option. To erase the shading, press 2nd PRGM ENTER to invoke the **ClrDraw** command on the Draw menu. The graph is then redrawn without the shading.

Graphing Derivative to Find Critical Points

In calculus, you need to graph the derivative of a function in order to find its critical points. Don't worry! Your calculator will help you; just follow these steps:

1. **Enter your functions in the Y= editor.**

2. **Use the ▶ ◀ ▲ ▼ keys to place your cursor in an open equation in the Y= editor.**

3. **Press MATH 8 to access the nDeriv template.**

4. **Press X,T,Θ,n, then press ALPHA TRACE and choose your function, then press ▶ X,T,Θ,n.**

   See the first screen in Figure 11-10. This is a sneaky move. You aren't using the **nDeriv(** template to take the derivative at a specific *x*-value. Instead, by taking the derivative at $x = x$, you are taking the derivative at all points where the function is defined.

Figure 11-10: Graphing the derivative to find the critical points.

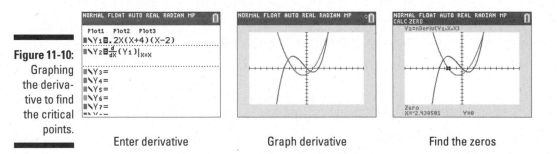

Enter derivative   Graph derivative   Find the zeros

5. **Press GRAPH to display the graph of your function and the derivative of the function.**

   See the second screen in Figure 11-10.

6. Press 2nd TRACE to access the Calculate menu.

7. Press 2 to select the zero option.

8. **If necessary, repeatedly press the ▲▼ keys until the derivative function appears in the border at the top of the screen.**

   Don't forget to do this step! (I had the original function selected the first time I tried it.)

9. **Enter the Lower and Upper Bounds, then a guess and press ENTER.**

   This is illustrated in the third screen in Figure 11-10. See the previous section in this chapter for the steps to find the zeros of a function.

Solving Equations by Graphing

There are a number of different ways to solve an equation by graphing. Next, I show you a method that I am particularly fond of. The basic idea is to set the equation equal to zero, graph it, and find the zeros. The method I show you has a twist that I think you will enjoy.

To solve the equation, $\sqrt{(x+2)} - 3 = 3 - x$, follow these steps:

1. **Set your equation equal to zero.**

   I subtracted 3 and added x to both sides:

   $$\sqrt{(x+2)} - 3 - 3 + x = 0$$

2. **Press Y= and enter one side of your equation in Y$_1$ and enter the other side of your equation in Y$_2$.**

   See the first screen in Figure 11-11. Notice, one of your functions is a horizontal line on the x-axis. I think it is easier to find intersection points than it is to find zeros. You might be wondering what the twist is with this method? When I find an intersection point of the two graphs, I have effectively found a zero of the equation! Pretty nifty, huh?

Figure 11-11:
Solving
equations
by graphing.

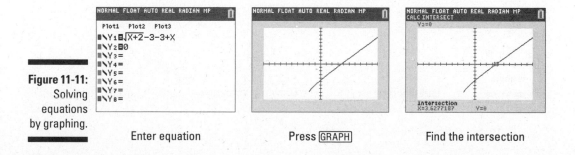

Enter equation   Press GRAPH   Find the intersection

3. **Press** GRAPH **and graph your functions in a viewing window that contains the intersection points of the two functions.**

   See the second screen in Figure 11-11. I like having a really nice window for this, but you don't have to. You may need to zoom out by pressing ZOOM 3 and pressing ENTER to make sure you have all the intersection points in the graphing window.

4. **Press** 2nd TRACE 5 ENTER ENTER ENTER **to find one of the intersection point(s) of the two graphed functions.**

   See the third screen in Figure 11-11. If there is more than one intersection point, you must press 2nd TRACE 5 ENTER ENTER and use the ▶ ◀ keys to navigate near the other intersection point. Press ENTER to make your guess.

You find your solution in the border at the bottom of the screen beneath the word "Intersection."

Drawing the Inverse of a Function

The big idea of inverse function is that x and y switch places. Your calculator has a built-in feature that enables you to "draw" the inverse of a function. Essentially, the calculator is "graphing" (not drawing) the inverse of the function. However, unlike a graph, you can't perform a trace or any other type of function evaluation on the drawn inverse. Another reason this term is used may be that the drawn inverse need not be a function.

Follow these steps to draw the inverse of a function:

1. **Enter your functions in the Y= editor.**

   See the first screen in Figure 11-12. I entered $Y_1 = e^x$. This function has a mathematically famous inverse, $f^{-1}(x) = \ln(x)$.

Figure 11-12:
Draw the inverse of a function.

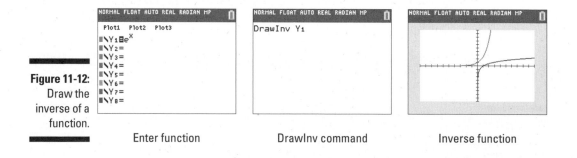

Enter function   DrawInv command   Inverse function

2. Press 2nd MODE to access the Home screen.

3. Press 2nd PRGM 8 to insert the DrawInv function

4. Press ALPHA TRACE and choose the name of the function you entered.

   See the second screen in Figure 11-12.

5. Press ENTER to display the graph of your function and draw the inverse of your function.

   See the third screen in Figure 11-12.

Enter $Y_2 = \ln(x)$ to double-check that the inverse your calculator drew is the natural log function. In the Y= editor, use the ◄ key and press ENTER to change the line style to ⸙. Then press GRAPH and enjoy the show!

Chapter 12

Graphing Inequalities

. . . .

In This Chapter

▶ Graphing one-variable inequalities

▶ Starting and quitting the Inequality app

▶ Entering and graphing inequalities

▶ Shading intersections and unions

▶ Storing data points

▶ Solving linear programming problems

. . . .

*W*ith the Inequality app that comes preloaded on the TI-84 Plus family of graphing calculators, you can graph functions and inequalities of the form $y \leq f(x)$, $y < f(x)$, $y \geq f(x)$, and $y > f(x)$. You can even graph and shade regions formed by the union or intersection of several inequalities. You can also use this app to solve linear programming problems. If you don't know what linear programming is, see the explanation in the linear programming section in this chapter.

Graphing One-Variable Inequalities

Sometimes, a product is used for tasks that it was not originally designed to accomplish. Play-Doh was originally meant to be a cleaner before it became a hit with kids everywhere. Your calculator was not made to graph inequalities on a number line, but it can be used to accomplish that task.

The reason your calculator is able to perform a task that it was not designed for is the Boolean logic your calculator uses to evaluate statements. If you read Chapter 7, you may remember that your calculator uses truth values: 1 = *True* and 0 = *False*.

When you enter a statement like **1 – X > 3**, your calculator figures out where the statement is true and returns a 1, and where the statement is false returns the value of 0. Why not use this to your advantage in a graphing environment?

Follow these steps to graph a one-variable inequality (as if graphing on a number line) on your calculator:

1. **Press Y= and enter the entire inequality.**

   See the first screen in Figure 12-1. I entered: $Y_1 = 1 - X > 3$.

   Press 2nd MATH to enter an inequality from the Test menu.

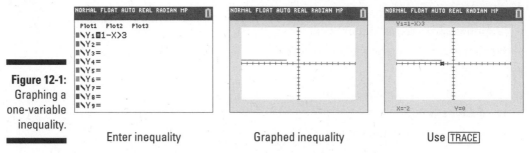

Figure 12-1: Graphing a one-variable inequality.

Enter inequality   Graphed inequality   Use TRACE

2. **Press ZOOM 6 to graph the one-variable inequality.**

   See the second screen in Figure 12-1. When the *y*-value of the graph is 1, the inequality is true. When the *y*-value of the graph is 0, the inequality is false.

   Your graph looks like a number line. The number line is slightly above the *x*-axis in a similar way that my students like to draw a number line on their homework. What is the only thing that is missing? Is the point at $x = -2$ opened or closed? See the next step for a method of checking the truth value at $x = -2$.

3. **Press TRACE and enter an *x*-value you would like to check.**

   See the third screen in Figure 12-1. I entered **−2**, which yielded a *y*-value of 0 (which means false). I can safely conclude the point is open at $x = -2$ and the solution inequality is $x < -2$.

   Press 2nd GRAPH to display the table. This is a really interesting way of looking at the truth values (of 1 and 0) that are returned for the inequality you enter.

This technique works for compound inequalities as well. The only drawback is that it can be difficult to determine the exact value where the graph begins.

The steps for entering compound inequalities are exactly the same as entering one-variable inequalities. In the first screen in Figure 12-2, I entered an "or" inequality for Y_1 and I entered an "and" inequality for Y_2. I graphed Y_1 in the second screen in Figure 12-2 and I graphed Y_2 in the third screen in Figure 12-2.

REMEMBER

Press 2nd MATH ▶ to insert "and" or "or" from the Logic menu.

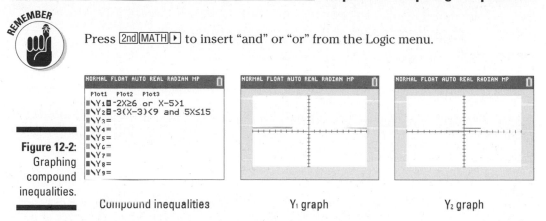

Figure 12-2:
Graphing
compound
inequalities.

Compound inequalities   Y₁ graph   Y₂ graph

Starting Inequality Graphing

You're probably used to using apps with some of the other technologies you own. The Inequality app is a powerful mathematics tool. To start the Inequality app, press APPS. See the first screen in Figure 12-3. Then, press ALPHA x^2; if necessary, use ▼ to move the cursor to the **Inequalz** app, and press ENTER to select the app. In the list of apps, this app is titled, Inequalz. However, the official name of the app is the Inequality Graphing app. After choosing the app, you are confronted with one of the last two screens shown in Figure 12-3.

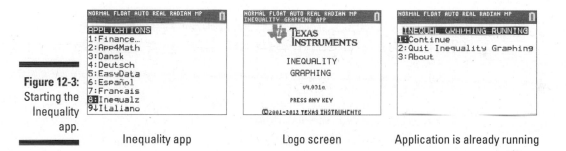

Figure 12-3:
Starting the
Inequality
app.

Inequality app   Logo screen   Application is already running

If no other apps are running, you see the second screen in Figure 12-3. Press any key to enter the Inequality Graphing app. If Inequality Graphing is already running, you see the third screen in Figure 12-3. Press 1 to re-enter the app.

After you enter Inequality Graphing, you are placed in the Y= editor so that you can enter functions and inequalities. The functions previously housed in this editor appear on the screen along with the inequality symbols at the bottom of the screen, as illustrated in the first screen in Figure 12-4. If you move the cursor so that it is not on an equal sign, the inequality symbols at the bottom of the screen vanish, as in the second image in this figure.

NORMAL FLOAT AUTO REAL RADIAN MP
SELECT RELATION-PRESS ALPHA F1-F5

X= Plot1 Plot2 Plot3   QUIT-APP
■\Y₁▉sin(X+π∕2)
■\Y₂=
■\Y₃=
■\Y₄=
■\Y₅=
■\Y₆=
■\Y₇=
■\Y₈=

    =   <   ≤   >   ≥

NORMAL FLOAT AUTO REAL RADIAN MP
INEQUALITY GRAPHING APP

X= Plot1 Plot2 Plot3   QUIT-APP
■\Y₁▉sin(X+π∕2)
■\Y₂=
■\Y₃=
■\Y₄=
■\Y₅=
■\Y₆=
■\Y₇=
■\Y₈=
■\Y₉=

Figure 12-4: The Y= editor when Inequality Graphing is running.

Cursor is on equal sign   Cursor is not on equal sign

REMEMBER

On the TI-84 Plus C, Context Help in the border at the top of the screen says, "Select Relation – Press Alpha *f1–f5*." If you forget a step while using this app, use the Context Help to guide you!

Entering Functions and Inequalities

The Inequality Graphing app can graph functions and inequalities of the form $y = f(x)$, $y < f(x)$, $y \le f(x)$, $y > f(x)$, and $y \ge f(x)$. Such functions and inequalities are defined in the Y= editor. The app can also graph equalities and inequalities of the form $x = N$, $x < N$, $x \le N$, $x > N$, and $x \ge N$, provided that N is a number. These equalities and inequalities are defined in the X= editor. Using these editors is explained in the following sections.

Entering inequalities in the Y= editor

To define a function or inequality of the form $y = f(x)$, $y < f(x)$, $y \le f(x)$, $y > f(x)$, and $y \ge f(x)$, follow these steps:

1. **Press Y= to access the Y= editor.**

   TIP

   To erase any unwanted functions or inequalities from the Y= editor, use the arrow keys to place the cursor after the equality or inequality symbol in the definition of the unwanted function or inequality and press CLEAR.

2. **Use the arrow keys to place the cursor on the sign (=, <, ≤, >, or ≥) of the function or inequality you are defining.**

3. **Press ALPHA and press the key under the appropriate equality or inequality symbol.**

   To get the first screen in Figure 12-5, I pressed ALPHA ZOOM to enter ≤. I pressed ZOOM because that is the soft key under the ≤ symbol appearing in the on-screen prompt at the bottom of the screen.

The equality and inequality symbols appear at the bottom of the screen only when the cursor is on the equality or inequality symbol appearing to the right of one of the functions Y_1 through Y_9 or Y_0.

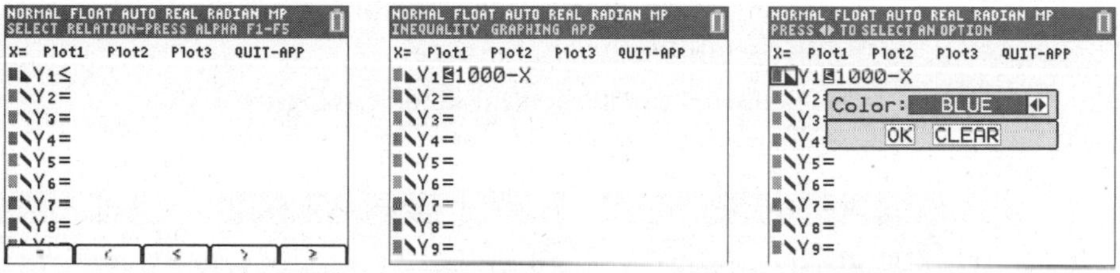

Figure 12-5:
Defining
inequalities
in the
Y= editor
and chang-
ing colors.

Place cursor on equal sign   Enter inequality symbol   Define inequality

4. Press ▷ and enter the definition of the function or inequality.

The definition of the function or inequality is entered the same way you enter the definition of a function, as explained in Chapter 9. The second screen in Figure 12-5 shows that the inequality $y \le 1000 - x$ is defined in Y_1.

After defining a function or inequality, you can change the inequality sign in this definition by following Steps 2 and 3.

Changing the color of inequalities

The graphing style of an inequality is determined by the Inequality Graphing app and cannot be changed. However, the color of the inequality graph can easily be changed. Use ◁ to move your cursor to the far left in the Y= editor and press ENTER. A color spinner menu as shown in the third screen in Figure 12-5 can be changed by using ▷ and ◁ keys.

When you exit the Inequality Graphing app, all inequality signs in the Y= editor are converted to equal signs and the original inequality sign is not reinstated the next time you run the app.

Entering inequalities in the X= editor

Equalities and inequalities of the form $x = N$, $x < N$, $x \le N$, $x > N$, and $x \ge N$ (where N is a number) are defined in the X= editor the same way inequalities are defined in the Y= editor, as explained in the preceding section. To access the X= editor, follow these steps:

1. **If you are not currently in the Y= editor, press** Y= **to get there.**

2. **Repeatedly press** ▲ **until the cursor is on X= in the upper-left corner of the Y= editor.**

   This is illustrated in the first screen in Figure 12-6.

Figure 12-6:
Defining
inequalities
in the
X= editor.

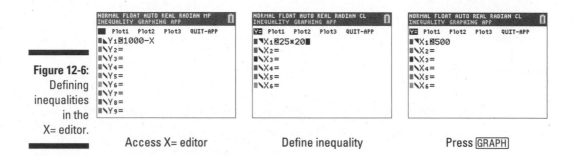

Access X= editor   Define inequality   Press GRAPH

3. **Press** ENTER **to access the X= editor.**

The *number* N in the inequality can be entered as an arithmetic expression, as illustrated in the second screen in Figure 12-6. That expression is evaluated by the calculator after you press ENTER, ▲, or ▼, as illustrated in the third screen in Figure 12-6.

To return to the Y= editor from the X= editor, repeatedly press ▲ until the cursor is in the upper-left corner of the screen, and then press ENTER.

When you exit the Inequality Graphing app, all entries made in the X= editor are erased.

Graphing Inequalities

As with functions, if the inequality sign in the definition of the inequality is highlighted, then that inequality will be graphed; if it isn't highlighted, it won't be graphed. The first screen in Figure 12-7 shows that the first two inequalities will be graphed, but the third won't. To change the highlighted status of an inequality sign, place the cursor on that sign and press ENTER.

You graph the inequalities defined in the Y= and X= editors the same way you graph a normal function. You either define the variables in the Window editor (as explained in Chapter 9) and press GRAPH or use one of the Zoom commands explained in Chapter 10. When the graph is displayed, you see three options at the bottom of the graph screen, as in the third screen in Figure 12-7. These options are explained in the next section.

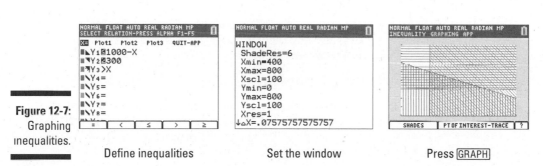

Figure 12-7: Graphing inequalities.

Define inequalities   Set the window   Press GRAPH

The Inequality Graphing app adds the **ShadeRes** variable to the Window editor, as illustrated in the second screen in Figure 12-7. This variable can be set to any integer 3 through 8. A setting of 3 places the lines in the shaded graph close together and 8 places them far apart. The default value is 6 as shown in the second screen in Figure 12-7, and this setting works just fine for most graphs, as illustrated in the third screen in Figure 12-7.

Exploring a Graph

When the graph is displayed, three options appear at the bottom of the screen, as illustrated in the third picture in Figure 12-7. The **Shades** option redraws the graph shading only the union or intersection of the regions, the **Pt of Interest-Trace** option traces the points of intersection appearing in the graph, and the **?** option displays the rudimentary Help screen shown in Figure 12-8. In this section, I explain how to use the **Shades** and **Pt of Interest-Trace** options, and I also explain what other options are available for exploring and investigating your graph.

Figure 12-8: The Inequality Graphing Help screen.

To hide the three options at the bottom of the graph screen, press ENTER; to redisplay them, press GRAPH.

Shading unions and intersections

The graph in the third picture in Figure 12-7 is pretty cluttered, as is the case with most graphs of more than two inequalities. The **Shades** option gets rid of the clutter by shading only the union or the intersection of the regions. To accomplish this, follow these steps:

1. **Press ALPHA Y= or ALPHA WINDOW to display the Shades menu.**

   Because the **Shades** option at the bottom of the graph screen is above the F1 and F2 function keys on the calculator, pressing either ALPHA Y= to select F1 or ALPHA WINDOW to select F2 produces the **Shades** menu, as illustrated in the first screen in Figure 12-9.

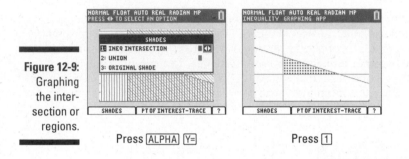

Figure 12-9:
Graphing the inter-section or regions.

Press ALPHA Y=   Press 1

2. **Use the ▶◀ keys to change the color spinner.**

   Choose from 15 available colors.

3. **Press the number of the option you want.**

   I chose 1.

   The new graph takes a little time to graph, but eventually appears on the screen, as illustrated in the second screen in Figure 12-9.

After graphing the union or intersection of the regions in your graph, you can redisplay the original shading of the graph by selecting the third option in the **Shades** menu.

Finding the points of intersection

The **Pt of Intersection-Trace** option is used to find the points of intersection appearing on the graph screen. When the calculator finds such a point, you can store the *x*- and *y*-coordinates of that point in the calculator. This is quite handy when solving linear programming problems, as explained later in this chapter. To find and store the points of intersection in an inequality graph, follow these steps:

1. **Press** ALPHA ZOOM **or** ALPHA TRACE **to select the Pt of Intersection-Trace option.**

   Because the **Pt of Intersection-Trace** option at the bottom of the graph screen is above the F3 and F4 function keys on the calculator, this option can be selected by pressing either ALPHA ZOOM to select F3 or ALPHA TRACE to select F4.

   After selecting this option, the cursor moves to one point of intersection and the coordinates of that point are displayed at the bottom of the screen, as illustrated in the first screen of Figure 12-10. In the upper-left corner of the screen, you see the names of the intersecting inequalities.

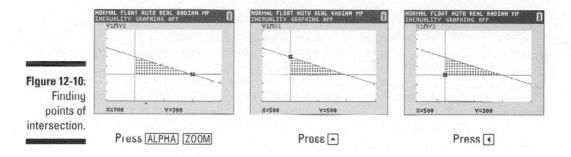

Figure 12-10:
Finding
points of
intersection.

Press ALPHA ZOOM   Press ▲   Press ◄

2. **Press** STO▶ **to store the coordinates of the point of intersection.**

   If you don't need to store these coordinates, you can skip this step. If you do press STO▶, you get a message saying, "Point appended to (∟NEQX, ∟INEQY)," as shown in the first screen in Figure 12-11. This tells you that the *x*-coordinate is stored in the list named INEQX and the *y*-coordinate is stored in list INEQY. Accessing, using, and managing these lists is explained later in this chapter in the section, "Storing Data Points." Press ENTER to get rid of the message and return to the graph.

   If the point is already stored in the calculator, you get the "Duplicate Point" message as shown in the second screen in Figure 12-11. Press ENTER to get rid of the message. The Inequality Graphing app will not store the point a second time.

3. **Use the arrow keys to move to the next point of intersection and, if you desire, press** STO▶ **to store its coordinates.**

   Pressing ◄ or ▶ moves the cursor to the next point of intersection on the graph of the left inequality in the upper-left corner of the screen. This is illustrated in the first and second screens in Figure 12-10. In the first screen in this figure, the left inequality in the upper-left corner of the screen is Y1. After I pressed ▲, the cursor jumped to the other point of intersection on this line, as illustrated in the second screen. Because this line has only two points of intersection, if I were to press ▲ again, the cursor would go back to the point of intersection in the first picture in this figure.

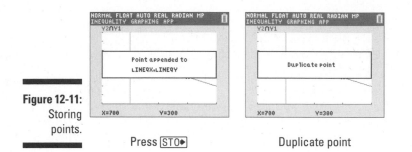

Figure 12-11:
Storing
points.

Press STO▶   Duplicate point

Pressing ▲ or ▼ changes the left inequality in the upper-left corner of the screen. This may or may not move the cursor. If the cursor doesn't move, press ◄ or ► to move the cursor to the next point of intersection. For example, to get from the second to the third screen in Figure 12-10, I pressed ◄ to change the left inequality to X1. But this gave me the intersection same point of intersection. So I pressed ◄ to get the other point of intersection on line X1, as illustrated in the third screen in Figure 12-10.

If **Pt of Intersection-Trace** is not able to find all points of intersection appearing on the graph screen, the "Pt of Intersection-Trace is not perfect" sidebar in this chapter gives you a solution to this problem.

4. **Press CLEAR when you are finished using Pt of Intersection-Trace.**

Pt of Intersection-Trace is not perfect

The Pt of Intersection-Trace feature found in the Inequality Graphing app has no trouble finding the points of intersection of linear inequalities. However, if your inequalities are not linear, **Pt of Intersection-Trace** may not be able to find all points of intersection. But this isn't a problem, because you can always use the **Intersect** tool in the **CALC** menu to find the points of intersection that **Pt of Intersection-Trace** couldn't find. And the points found by the **Intersect** tool can be stored in the calculator the same way that points found by **Pt of Intersection-Trace** are stored. Using the **Intersect** tool to find points of intersection is explained in Chapter 11. The "Storing Data Points" section, later in this chapter, explains how to store points of intersection found using the **Intersect** tool.

Other ways to explore a graph

All the commands and features described in Chapters 9, 10, and 11 that are available for graphing and exploring normal functions are also available when graphing and exploring inequalities. For example, you can split the screen and display a graph and a table, you can zoom in on a graph, you can trace a graph, you can find the coordinates of maximum and minimum points, and you can do much, much more. Many of these commands are housed in the **Zoom** and **CALC** menus. Using these commands and features is explained in Chapters 9, 10, and 11.

Storing Data Points

When you start Inequality Graphing for the first time, the app creates two lists, INEQX and INEQY, to house the *x*- and *y*-coordinates of data points that you store in the calculator. When you exit Inequality Graphing, these data lists are not deleted from the calculator. So when you start the app again at a later time, any data previously stored in these lists will still be there, provided that you didn't delete the lists from the memory of the calculator.

The final section in this chapter, "Solving Linear Programming Problems," gives a real-world example of why you would want to store data points in lists. In this section, I explain how to store data points in these lists, clear the contents of these lists, and view the data in the lists. Chapter 17 gives you a more detailed explanation of dealing with data lists. Among other things, Chapter 17 tells you how to manually enter or edit data in a list, how to delete data from a list, and how to sort data. Chapter 17 also explains how to delete a data list from the calculator's memory. But if you do delete the INEQX and INEQY data lists, the next time you start Inequality Graphing, the app will re-create these lists. So why bother deleting them?

Clearing the INEQX and INEQY lists

When the Inequality Graphing app stores a data point, it appends that point to the other points already stored in the INEQX and INEQY lists. When you exit Inequality Graphing, the app does not clear the contents of these lists. So if you are graphing a new set of inequalities and want to store data points associated with the graph, it's a good idea to clear the old data points from these lists.

To clear the contents of the INEQX and INEQY lists, press GRAPH ALPHA CLEAR 2. The graph screen appears and the INEQX and INEQY lists remain in the calculator as empty data lists.

If an inequality graph is displayed on the screen, simply press ALPHA CLEAR 2 to clear the INEQX and INEQY data lists. If any other screen is displayed, such as the Y= editor, you must press GRAPH ALPHA CLEAR 2 to clear these lists.

Storing points in *INEQX* and *INEQY*

The Inequality Graphing app can store any data point whose coordinates appear at the bottom of the graph screen. Such points are found using **Trace**, **Pt of Intersection-Trace**, or any of the first five tools in the **CALC** menu.

A data point can be stored in the calculator only when the cursor is on that point and the coordinates of the cursor location appear at the bottom of the graph screen. When this is the case, press STO▸ to store the coordinates of the data point. You usually get a message saying, "Point appended to (∟INEQX, ∟INEQY)" or "Duplicate point." The first message tells you that the x-coordinate of your data point is stored in the list named INEQX and the y-coordinate is stored in list INEQY. Press ENTER to get rid of the message and return to the graph. If the point is already stored in the calculator, you get the "Duplicate Point" message. Press ENTER to get rid of the message. The Inequality Graphing app will not store the point a second time.

Viewing stored data

Press STAT 1 to view the data stored in lists INEQX and INEQY, as illustrated in Figure 12-12. If the Stat List editor already contains 19 or 20 lists, then there is not enough room for the editor to display these two lists. This situation is remedied by deleting a few lists from the Stat List editor and then recalling the INEQX and INEQY lists. How to do this is explained in Chapter 17.

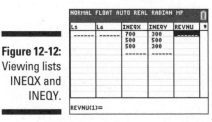

Figure 12-12: Viewing lists INEQX and INEQY.

Solving Linear Programming Problems

Linear programming is a method for finding the maximum or minimum value of a multivariable function that is constrained by a system of inequalities. The following example should help you understand this rather technical definition of linear programming.

> *A chocolate company sells real and imitation chocolate chips to a local cookie factory. On any given day, the cookie factory needs at least 500 pounds of real chocolate chips and at least 300 pounds of imitation chocolate chips. The real chocolate chips sell for $1.25 a pound and the imitation chocolate chips sell for $0.75 a pound. If the truck that takes the chocolate chips to the cookie factory can carry at most 1,000 pounds of chocolate chips, how many pounds of each kind of chocolate chips should the chocolate company ship to the cookie factory in order to maximize its revenue?*

In this example, the chocolate factory's revenue is the function revenue = $1.25x + 0.75y$, where x is the number of pounds of real chocolate chips and y is the number of pounds of imitation chocolate chips that the chocolate company ships to the cookie factory. The constraints stated in this example are: $500 \leq x$, $300 \leq y$, and $x + y \leq 1000$. In other words, this example asks you to find the maximum value of revenue = $1.25x + 0.75y$ subject to the system of constraints $x \geq 500$, $y \geq 300$, and $y < 1000 - x$.

How do you solve a linear programming problem? The following theorem gives the answer.

> **Linear Programming Theorem:** If an optimum (maximum or minimum) value of a function constrained by a system of inequalities exists, then that optimum value occurs at one or more of the vertices of the region defined by the constraining system of inequalities.

This theorem tells you to evaluate the function at the points of intersection of the constraining system of inequalities. The smallest value found is the minimum value of the function and the largest is its maximum value. To get the Inequality app to help you solve a linear programming problem, follow these steps:

1. **Graph the system of constraints.**

   How to do this is explained earlier in this chapter. For the example at the beginning of this section, the graph of the system of constraints ($x \geq 500$, $y \geq 300$, and $y \leq 1000 - x$) appears in the third screen in Figure 12-7.

2. Graph the intersection of the regions in the graph.

How to do this is explained earlier in this chapter. For the example at the beginning of this section, the graph of the intersection appears in the second screen in Figure 12-9.

3. Find and store the points of intersection in the graph.

How to do this is explained earlier in this chapter. For the example at the beginning of this section, the process of finding the points of intersection is illustrated in Figure 12-10.

4. Display the stored points of intersection.

How to do this is explained earlier in this chapter. For the example at the beginning of this section, the stored points of intersection appear in Figure 12-12.

5. Create a list to the right of list INEQY and give it a name.

The name you give the list should describe the function in the linear programming problem. In the next step, this function is evaluated at the stored points of intersection.

If an empty, unnamed list does not appear to the right of list INEQY, place the cursor in the heading of the third column and press 2nd DEL to insert a blank column, enter a name, and then press ENTER. If an empty, unnamed list does appear in the third column, place the cursor in the heading of that column, enter a name, and then press ENTER. Should you need it, a more detailed explanation of creating and naming lists can be found in Chapter 17.

For the example at the beginning of this section, the created list is named REVNU for revenue, as illustrated in Figure 12-12.

6. Use a formula to define the entries in the new list.

The formula you enter is the formula that defines the function you want to optimize. In the example at the beginning of this section, that formula is $1.25x + 0.75y$, the definition of the revenue function. Because x is housed in list INEQX and y in INEQY, this formula is entered into the calculator as $1.25*\llcorner$INEQX$ + 0.75*\llcorner$INEQY.

To use a formula to define a list, place the cursor on the name of the list in the column heading. Because formulas should be surrounded by quotes, press ALPHA + to enter the first quotation mark. Then enter the formula. To enter the name of a list, such as $\llcorner$INEQX, press 2nd STAT to display a list of the names of the lists in the Stat List editor. Repeatedly press ⏷ to highlight the number to the left of the list and press ENTER to insert the name of the list in your formula. After entering the formula, press ALPHA + to enter the closing quotation mark.

TIP

When you define a list, if you don't use quotes around the formula, it will still generate a list. However, if you change the values in the list, other lists will not update accordingly. For that reason, it is a good idea to use quotes around a formula when defining a list.

7. **Press** ENTER **to evaluate the function at the points of intersection of the constraining system of inequalities.**

According to the Linear Programming Theorem, if the function has a maximum and/or minimum value, those values appear in the list you just created. As illustrated in Figure 12-13, the chocolate factory in the example at the beginning of this section can maximize its revenue by shipping 700 pounds of real chocolate chips and 300 pounds of imitation chocolate chips.

Figure 12-13: Solution to a linear programming problem.

	NORMAL FLOAT AUTO REAL RADIAN MP				
L5	L6	INEQX	INEQY	REVNUA	9
-------	-------	700	300	1100	
		500	500	1000	
		500	300	850	
		-------	-------	-------	

REVNU(4)=

Quitting Inequality Graphing

Most of the time, you don't even know the Inequality Graphing app is running unless you are actively using the app or unless you place the cursor on an equal sign in the Y= editor to display the inequality symbols, as in the first screen in Figure 12-14. How to quit the app is about as puzzling as knowing whether or not it is running.

Figure 12-14: Quitting the Inequality app.

Quit-App   Press 2

To quit (exit) this app on the TI-84 Plus C, press ⎡Y=⎤ to access the Y= editor. Use the ⎡▶⎤⎡◀⎤⎡▲⎤⎡▼⎤ keys to move your cursor to "Quit-App," located in the top-right part of the screen as shown in the first image in Figure 12-14. Press ⎡ENTER⎤ and the second screen in Figure 12-14 appears. Press ⎡2⎤ to quit the app. On the TI-84 Plus, press ⎡APPS⎤ choose Inequalz and press ⎡2⎤ to quit the app. I guess that wasn't so bad after all.

Chapter 13

Graphing Parametric Equations

· ·

In This Chapter

▶ Changing the mode and window of your calculator

▶ Entering and graphing parametric equations

▶ Using Trace to evaluate parametric equations

▶ Viewing the table of a parametric graph

▶ Finding the derivative of parametric equations

· ·

*P*arametric equations are used in Pre-calculus and Physics classes as a convenient way to define x and y in terms of a third variable, T. If you are familiar with the graphing function on your calculator, then parametric equations shouldn't be too much of a challenge for you. In this chapter, you find tips and steps that should make graphing parametric equations something that you look forward to.

Anything that can be graphed in Function mode can also be graphed as a set of parametric equations. Using parametric equations enables you to investigate horizontal distance, x, and vertical distance, y, with respect to time, T. This adds a new dimension to your graph! The direction a point is moving is an important part of graphing parametric equations. Fortunately, your calculator does a good job of letting you see the direction of motion as the graph forms.

Changing the Mode

You can't begin graphing parametric equations until you change the mode of your calculator. Follow these steps to change the mode of your calculator:

1. **Press [MODE] and put the calculator in Parametric mode.**

   To highlight an item in the Mode menu, use the [▶][◀][▲][▼] keys to place the cursor on the item, and then press [ENTER]. Highlight **PARAMETRIC** in the fifth line to put the calculator in Parametric mode. See the first screen in Figure 13-1.

It is usually a good idea to put your calculator in Radian mode when working with parametric equations.

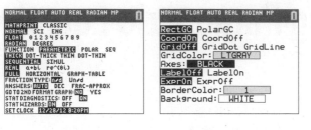

Figure 13-1:
Changing
the mode to
Parametric.

Parametric mode   Format menu

2. **Press 2nd ZOOM to access your Format menu.**

Make sure the ExprOn and CoordOn are both highlighted. These settings are helpful when performing a trace on your equations. See the second screen in Figure 13-1.

Selecting the Graph Style

All the functionality that you are used to having in Function mode is also available in Parametric mode. If you would like to customize your graph by changing the color or the line style, follow these steps:

1. **Press Y= to access the Y= editor.**

2. **Press ◄ to navigate your cursor to the left of the equal sign.**

See the first screen in Figure 13-2.

Figure 13-2:
Selecting
the Graph
Style.

Press ◄   Change color   Change line style

3. **Press** ENTER **and use the** ▷◁ **keys to change the color using the spin-ner menu.**

   See the second screen in Figure 13-2.

4. **Press** ENTER **and use the** ▷◁ **keys to change the line style using the spinner menu.**

   See the third screen in Figure 13-2.

5. **Press** ENTER **twice to make the changes effective.**

Entering Parametric Equations

If you are paying attention, then you may have noticed the Y= editor looks quite different than you may be used to. It seems like Y_1 has been replace with two equations, X_{1T} and Y_{1T}! Remember, the *x* and *y* variables are now defined in terms of a new parameter, T. When you press X,T,Θ,*n* in Parametric mode, a T appears instead of an *x*.

Usually, you are given a pair of parametric equations to graph with an interval for T. For this exercise, I use these parametric equations: $x(T) = 8\sin(T)$ and $y(T) = 4\cos(T)$, where $0 \le T \le 2\pi$.

1. **Press** Y= **to access the Y= editor.**

2. **Enter** $X_{1T} = 8\sin(T)$.

   Be sure to press X,T,Θ,*n* for T. See the first screen in Figure 13-3.

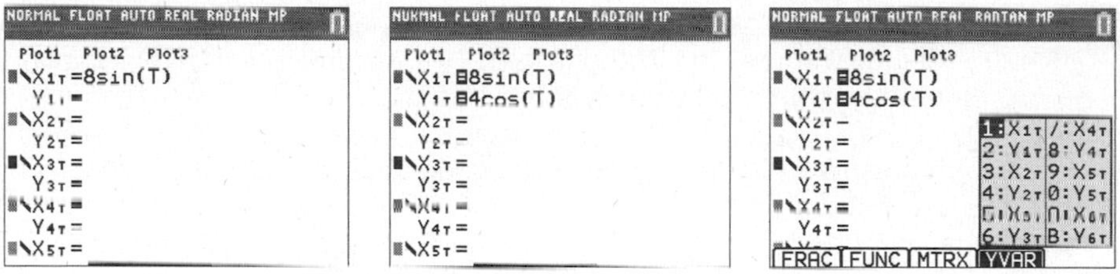

Figure 13-3:
Entering parametric equations.

| Enter X₁ₜ | Enter Y₁ₜ | Y-VAR menu |

3. **Enter** $Y_{1T} = 4\cos(T)$.

   See the second screen in Figure 13-3.

Press [ALPHA][TRACE] to access the Y-VAR menu. Your calculator has a customized Y-VAR menu so that you can enter variables like X_{1T} or Y_{1T} in your parametric equations. See the third screen in Figure 13-3.

Setting the Window

Setting the window in Parametric mode is a crucial step in graphing parametric equations. In fact, if my students are having trouble graphing parametric equations, it is usually because of the way they have set up their window. Specifically, three window settings tend to cause problems: Tmin, Tmax, and Tstep.

The interval for T was given in the problem $0 \le T \le 2\pi$. So, identifying Tmin and Tmax is pretty easy for this problem. This is going to sound strange, but changing the minimum and maximum values of T doesn't affect the viewing window of your graph. You would have to change the minimum and maximum values of X and Y to change the graphing window. What do the T values affect? The maximum and minimum T values affect how much of the graph you see. In Function mode, piecewise functions have a restricted domain so that you can only see a "piece" of the function. In Parametric mode, the T values can be restricted, which can make it difficult to predict what the "whole" graph would look like if the T values were not restricted to a certain interval.

How do you decide the size of Tstep? Tstep is the increment that your graph uses to plot each point in creating the graph you see on the screen. As a general rule of thumb, the smaller your step is, the more accurate your graph is going to be. The drawback is as the step gets smaller, your calculator takes longer to graph your parametric equations. As a general rule of thumb, the default value of the TStep is usually a good balance between graph accuracy and the time it takes to graph.

If you are in Radian mode, it is a good idea to set your Tstep as a π divided by a number.

Here are the steps to set your graphing window:

1. **Press [WINDOW] to access the window editor.**

   See the first screen in Figure 13-4.

2. **Change the value of Tmin and Tmax.**

   Remember, the interval for T values is $0 \le T \le 2\pi$. I entered 2π for Tmax and didn't press e as shown in the second screen in Figure 13-4.

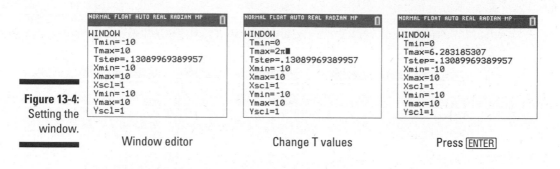

Figure 13-4:
Setting the
window.

Window editor   Change T values   Press ENTER

3. **Press** ENTER.

Notice, pressing ENTER evaluates 2π and the approximate value of
6.283185307 is displayed. See the third screen in Figure 13-4.

Graphing Parametric Equations

You have done all the heavy lifting; this step is easy. Before you press GRAPH,
make sure you watch the direction that your graph is created. Your calculator begins graphing by substituting the smallest T value in the interval. If
your Tstep is small enough, you should be able to see the graph develop.
When graphing parametric equations by hand, my students use arrows on
the graph to indicate the direction of motion.

Press GRAPH. See the graph in Figure 13-5.

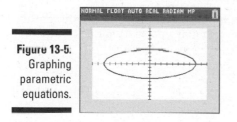

Figure 13-5:
Graphing
parametric
equations.

Using Zoom to Change the Window

If the graphing window is not to your liking, you can use any of the Zoom
commands described in Chapter 10. For example, if you are graphing the
parametric equations shown in the first screen in Figure 13-6, you may not be

happy with the graphing window shown in the second screen in Figure 13-6. Press $\boxed{\text{ZOOM}}\,\boxed{2}\,\boxed{\text{ENTER}}$ to zoom in as illustrated in the third screen in Figure 13-6.

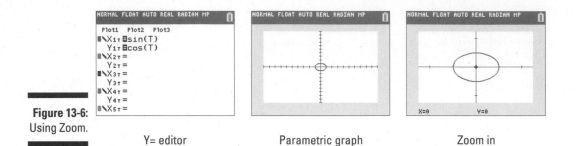

Figure 13-6:
Using Zoom.

Y= editor   Parametric graph   Zoom in

Using Trace to Evaluate a Parametric Equation

You are going to love using the Trace feature to evaluate parametric equations. I am impressed with how much information fits in the graph border around the graph screen. Remember, you are not tracing *x*-values as you do in Function mode. Follow these steps to evaluate a function at specific T values:

1. Press $\boxed{\text{TRACE}}$**.**

See the first screen in Figure 13-7. Your trace starts at the smallest T value in the interval defined in the Window editor. The values of X, Y, and T are all displayed in the border at the bottom of your graph screen.

The TI-84 Plus C displays functions and information in the border of the graph screen. The TI-84 Plus displays similar information directly on the graph screen.

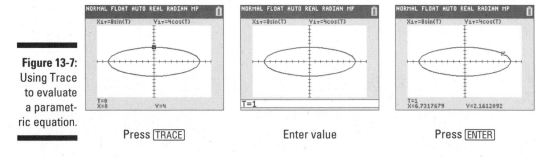

Figure 13-7:
Using Trace
to evaluate
a paramet-
ric equation.

Press $\boxed{\text{TRACE}}$ Enter value   Press $\boxed{\text{ENTER}}$

2. **Press ▶ to find the direction of motion of the parametric equations.**

   Pay attention to the direction of motion as you increase the value of T.

3. **Enter a specific T value.**

   After pressing TRACE, entering a number opens up an entry line in the border at the bottom of your graph screen, as shown in the second screen in Figure 13-7.

4. **Press ENTER.**

   See the result as shown in the third screen in Figure 13-7.

Viewing the Table of a Parametric Graph

It is easy to view the values of X, Y, and T all in one table. Press 2nd GRAPH to view the table as shown in the first screen in Figure 13-8.

Read the Context Help in the border at the top of the table, "Press + for ΔTbl." To change the table increment, press + and edit the value at the bottom of the screen as shown in the second screen in Figure 13-8.

Figure 13-8:
Viewing the table of a parametric graph.

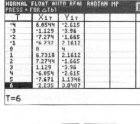

Press 2nd GRAPH

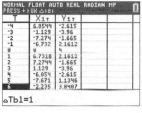

Table increment

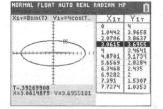

GRAPH-TABLE

Another option is to show a split screen with a graph and a table. Press MODE, use the ▶ ◀ ▲ ▼ keys to highlight GRAPH-TABLE and press ENTER. Press GRAPH to see the split screen. Using Trace in the Graph-Table mode automatically highlights the ordered pairs in the table, as shown in the third screen in Figure 13-8.

Taking the Derivative of Parametric Equations

If you need to take the derivative of parametric equations, follow these steps:

1. **Press** GRAPH.

2. **Press** 2nd ZOOM **to access the Calculate menu.**

   There are three options for derivatives when working in Parametric mode: dy/dx, dy/dt, and dx/dt. See the first screen in Figure 13-9.

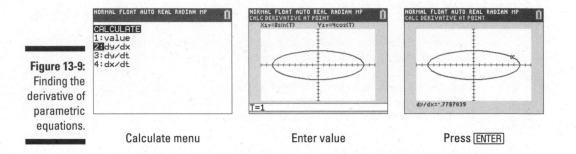

Figure 13-9: Finding the derivative of parametric equations.

Calculate menu   Enter value   Press ENTER

3. **Press** 2 **for dy/dx,** 3 **for dy/dt, or** 4 **for dx/dt.**

4. **Enter a specific T value where you want to find the derivative.**

   Entering a number opens up an entry line in the border at the bottom of your graph screen, as shown in the second screen in Figure 13-9.

5. **Press** ENTER.

   See the result in the border at the bottom of the graph screen, as shown in the third screen in Figure 13-9.

To find multiple derivatives, repeat Steps 2 through 5.

Chapter 14

Graphing Polar Equations

A polar coordinate system is used in Pre-calculus class as yet another way to define a point. Polar coordinates are of the form $(\boldsymbol{r}, \theta)$. The distance from the pole (similar to the origin) is called, r. The polar axis is a ray that extends from the pole (similar to the positive x-axis). A positive angle is measured in a counterclockwise direction from the polar axis to a line that connects the pole and a point. See Figure 14-1 for a visual of a polar coordinate.

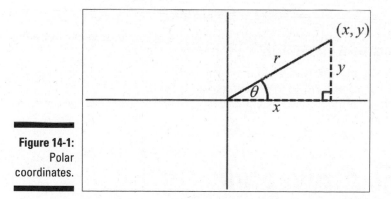

Figure 14-1:
Polar
coordinates.

Polar coordinates $(\boldsymbol{r}, \theta)$ can be converted to rectangular coordinates (x, y), as discussed in Chapter 9. The purpose of this chapter is to explain how to enter and graph polar equations. As you might imagine, things look a little different in Polar mode. For starters, the Y= editor could temporarily change its polar name to the r= editor. If you keep reading, you will get the hang of what I am referring to and you will be graphing polar equations in no time at all!

Changing the Mode

You can't begin graphing polar equations until you change the mode of your calculator. Follow these steps to change the mode of your calculator:

1. **Press** MODE **and put the calculator in Polar mode.**

   To highlight an item in the Mode menu, use the ▶◀▲▼ keys to place the cursor on the item, and then press ENTER. Highlight **POLAR** in the fifth line to put the calculator in Polar mode. See the first screen in Figure 14-2.

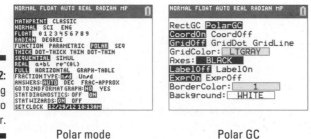

Figure 14-2:
Changing
the mode to
Polar.

Polar mode   Polar GC

Polar equations can be graphed in Radian or Degree mode; pay attention to the problem presented and set your mode accordingly. I set the mode to Radian.

2. **Press** 2nd ZOOM **to access your Format menu.**

   Make sure the ExprOn and CoordOn are both highlighted. These settings are helpful when performing a trace on your equations.

   Your have an important decision to make! Do you want your coordinates displayed in polar form (r, θ) or rectangular form (x, y)? I chose polar form by highlighting **Polar GC** and pressing ENTER as shown is the second screen in Figure 14-2.

Selecting the Graph Style

All the functionality that you are used to having in Function mode is also available in Polar mode. In Polar mode, you can create graphs that look like roses, so changing the color of your graph might be important to you (especially if you want a red rose.) If you would like to customize your graph by changing the color or the line style, follow these steps:

1. **Press** Y= **to access the Y= editor.**

2. **Press** ◄ **to navigate your cursor to the left of the equal sign.**

   See the first screen in Figure 14-3.

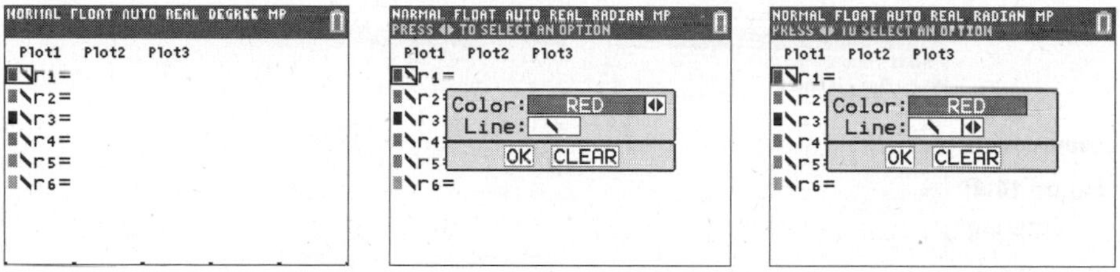

Figure 14-3:
Selecting
the Graph
Style.

Press ◄   Change color   Change line style

3. **Press** ENTER **and use** ► **and** ◄ **keys to change the color using the spinner menu.**

   See the second screen in Figure 14-3.

4. **Press** ENTER **and use the** ► **and** ◄ **keys to change the line style using the spinner menu.**

   See the third screen in Figure 14-3.

5. **Press** ENTER **twice to make the changes effective.**

Entering Polar Equations

You may have noticed the Y= editor looks a little different than you may be used to. Y_1 has been replaced with r_1. That is not all that has changed; when you press X,T,Θ,n in Polar mode, a θ appears instead of an x.

Polar graphs take on all sorts of interesting shapes: spirals, limaçons, cardioids, lemniscates, and roses, just to name a few. These graphs are usually symmetric over the polar axis or the vertical axis.

For this exercise, I use the polar equation: r = 4cos(6θ) with a range of $0 \leq \theta \leq 2\pi$. This polar equation forms a rose curve.

1. **Press** Y= **to access the Y= editor.**

2. **Enter** r_1 = 4cos(6θ).

Be sure to press $\boxed{\text{X,T,}\Theta\text{,}n}$ for r. See the first screen in Figure 14-4.

Press $\boxed{\text{ALPHA}}\boxed{\text{TRACE}}$ to access the Y-VAR menu. Your calculator has a customized Y-VAR menu so that you can save time by entering variables like r_1 or r_2 in your polar equations. See the second screen in Figure 14-4.

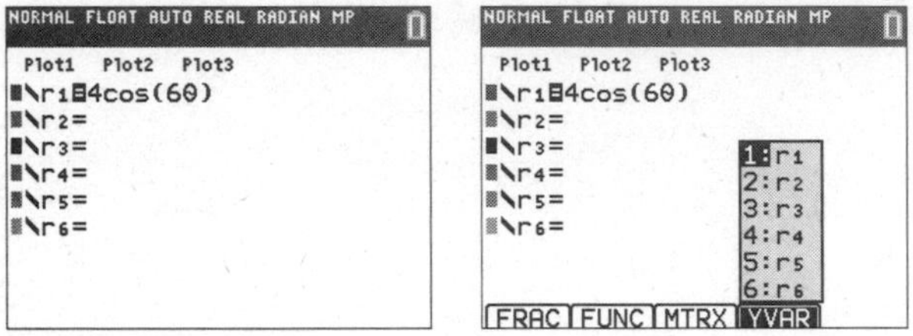

Figure 14-4:
Entering
polar
equations.

Enter r_1 Y-VAR menu

Setting the Window

Before graphing a polar graph, set your window. If your graph seems incomplete, it is probably due to the way you set your window variables. The variables that tend to cause problems are θmin, θmax, and θstep.

The range given in the problem is $0 \le \theta \le 2\pi$. It is easy to see that θmin=0 and θmax=2π. Even though these variables are part of the Window editor, they don't actually affect the viewing window of the graph on your calculator. You would have to change the minimum and maximum values of X and Y to change the graphing window. Does that seem strange? Maybe this explanation will help. In Function mode, piecewise functions have a restricted domain so that you can only see a "piece" of the function. In Polar mode, the range can be restricted, which can make it difficult to predict what the "whole" graph would look like if the θ values were not restricted to a certain interval. As a general rule of thumb, you should be able to see the whole graph if $0 \le \theta \le 2\pi$ in Radian mode, or $0 \le \theta \le 360$ in Degree mode.

θstep is the increment between θ values. When you graph a polar equation, your calculator evaluates r for each value of θ by increments of θstep to plot each point. Be careful! If you choose a θstep that is too large, your polar graph will not be accurate. If you choose a θstep that is too small, it will take a long time for your calculator to graph. In the ZStandard window, the default value for θstep is π/24 in Radian mode or 15 in Degree mode. In most cases, this is a good balance between graphing accuracy and the time it takes to graph.

Follow these steps to set the window for a Polar graph:

1. **Press WINDOW to access the Window editor.**

   See the first screen in Figure 14-5.

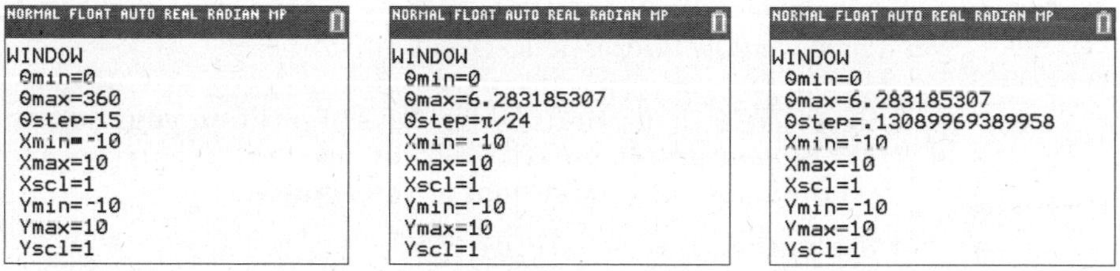
NORMAL FLOAT AUTO REAL RADIAN MP
WINDOW
θmin=0
θmax=360
θstep=15
Xmin=-10
Xmax=10
Xscl=1
Ymin=-10
Ymax=10
Yscl=1
NORMAL FLOAT AUTO REAL RADIAN MP
WINDOW
θmin=0
θmax=6.283185307
θstep=π/24
Xmin=-10
Xmax=10
Xscl=1
Ymin=-10
Ymax=10
Yscl=1
NORMAL FLOAT AUTO REAL RADIAN MP
WINDOW
θmin=0
θmax=6.283185307
θstep=.13089969389958
Xmin=-10
Xmax=10
Xscl=1
Ymin=-10
Ymax=10
Yscl=1

Window editor | Change θ values | Press ENTER

Figure 14-5: Setting the window.

2. **Change the value of θmin, θmax, and θstep.**

   Remember, the range for the problem is $0 \le \theta \le 2\pi$. I entered 2π for θmax, and pressed ENTER. I entered $\pi/24$ for θstep, and did not press ENTER as shown in the second screen in Figure 14-5.

3. **Press ENTER.**

   Notice, pressing ENTER evaluates $\pi/24$ and the approximate value of 0.13089969389958 is displayed. See the third screen in Figure 14-5.

Graphing Polar Equations

After you have done all the preparations, this step is easy. Before you press GRAPH, make sure you watch the direction as your graph is created. If your θstep is small enough, you should be able to see the graph develop.

Press GRAPH. See the graph in Figure 14-6.

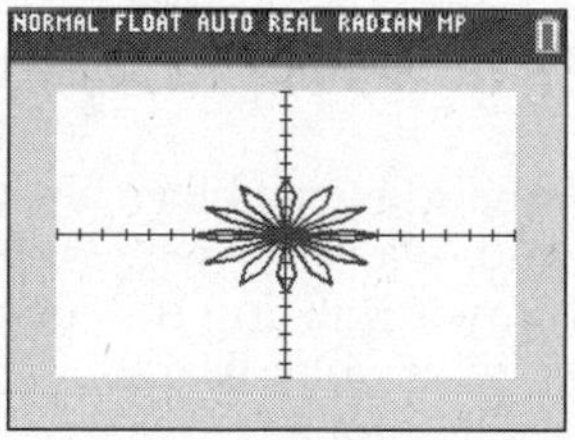
NORMAL FLOAT AUTO REAL RADIAN MP

Figure 14-6: Graphing polar equations.

Using Zoom to Change the Window

If the graphing window is not to your liking, you can use any of the Zoom commands described in Chapter 10. Here are the steps that I performed to get a nice window for the polar graph.

1. **Press** ZOOM 1, **use the** ▶◀▲▼ **keys to position your cursor above and left of your graph, press** ENTER, **and use the** ▶◀▲▼ **keys to reposition your cursor below and right of your graph.**

   See the first screen in Figure 14-7.

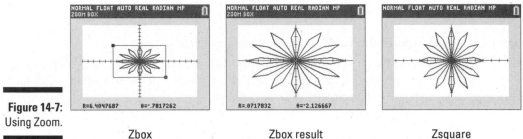

Figure 14-7:
Using Zoom.

Zbox   Zbox result   Zsquare

2. **Press** ENTER.

   See the result of ZBox in the second screen in Figure 14-7.

3. **Press** ZOOM 5.

   See the result of Zsquare in the third screen in Figure 14-7.

Sometimes you try a Zoom command and don't particularly like the result (I did this with zoom in). No worries! Press ZOOM ▶ ENTER to invoke the ZPrevious command and return to whatever zoom you last used.

Using Trace to Evaluate a Polar Equation

Using the Trace feature to evaluate polar equations can be easily managed. If your Format is set to **Polar GC**, you will find *r*-values when you trace the polar graph. If your Format is set to **Rect GC**, you will find the rectangular coordinates (*x*,*y*) of the points that make up your polar graph.

Follow these steps to evaluate a polar equation at specific θ values:

1. **Press** TRACE.

   See the first screen in Figure 14-8. Your trace starts at the θmin value as defined in the Window editor. The value of r is displayed in the border at the bottom of your graph screen.

   The TI-84 Plus C displays functions and information in the border of the graph screen. The TI-84 Plus displays similar information directly on the graph screen.

   After pressing TRACE, use ▶ and ◀ to investigate points at different θ values. Get trace crazy and press 2nd▶ or 2nd◀ to move five plotted points at a time! Try it! It's fun!

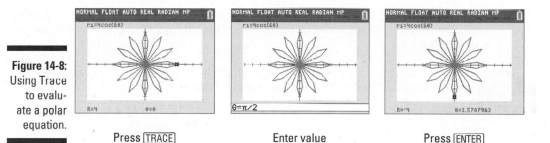

Figure 14-8:
Using Trace to evaluate a polar equation.

Press TRACE   Enter value   Press ENTER

2. **Enter a specific θ value.**

   After pressing TRACE, entering a number opens up an entry line in the border at the bottom of your graph screen, as shown in the second screen in Figure 14-8.

   If your calculator is in Radian mode, enter the angle, θ, in radians. If your calculator is in Degree mode, enter the angle, θ, in degrees.

3. **Press** ENTER.

   See the result as shown in the third screen in Figure 14-8.

Viewing the Table of a Polar Graph

It is easy to view the values of your variables all in one table. Press 2nd GRAPH to view the table as shown in the first screen in Figure 14-9.

Read the Context Help in the border at the top of the table, "Press + for ΔTbl." To change the table increment, press ⊞ and edit the value at the bottom of the screen as shown in the second screen in Figure 14-9.

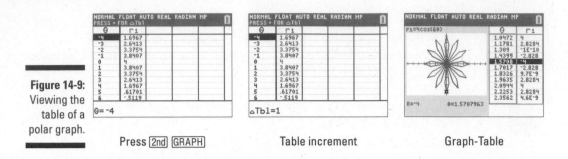

Figure 14-9:
Viewing the
table of a
polar graph.

Press [2nd] [GRAPH]   Table increment   Graph-Table

Another option is to show a split screen with a graph and a table. Press [MODE],
use the [▶][◀][▲][▼] keys to highlight GRAPH-TABLE, and press [ENTER]. Press
[GRAPH] to see the split screen. Using Trace in the Graph-Table mode automatically highlights the ordered pairs in the table, as shown in the third screen in
Figure 14-9.

When using Trace, if Format is set to **Rect GC**, you will see *X, Y,* displayed
on the graph screen. This means you can view *X, Y,* **r**, and θ all on the same
screen! Wow!

Taking the Derivative of Polar Equations

If you need to take the derivative of polar equations, follow these steps:

1. **Press [GRAPH].**

2. **Press [2nd][ZOOM] to access the Calculate menu.**

   There are two options for derivatives when working in Polar mode: dy/
   dx and dr/dθ. See the first screen in Figure 14-10.

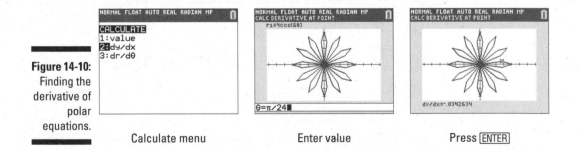

Figure 14-10:
Finding the
derivative of
polar
equations.

Calculate menu   Enter value   Press [ENTER]

3. **Press** ② **for dy/dx,** ③ **for dr/dθ.**

4. **Enter a specific θ value where you want to find the derivative.**

   Entering a number opens up an entry line in the border at the bottom of your graph screen, as shown in the second screen in Figure 14-10.

5. **Press** ⸢ENTER⸥.

   See the result in the border at the bottom of the graph screen, as shown in the third screen in Figure 14-10.

To find multiple derivatives, repeat Steps 2 through 5.

Chapter 15

Graphing Sequences

. . . .

In This Chapter

▶ Changing the mode and window of your calculator

▶ Entering and graphing sequences

▶ Using Trace to evaluate sequences

▶ Graphing a recursive sequence

▶ Graphing a web plot

▶ Graphing a phase plot

▶ Graphing partial sums of an infinite series

. . . .

Sequences make interesting graphs! I often find that using mathematics to look at data in graphical form helps me to better understand the data and what it represents. Of course, you need to learn how to set the mode and window and enter a sequence in your calculator before you can have the fun of graphing a sequence. If you want to challenge yourself to learn something new, read about graphing a web plot or a phase plot (the graphs really look cool).

Changing the Mode

You can't begin graphing sequences until you change the mode of your calculator. Follow these steps to change the mode of your calculator:

1. **Press** MODE **and put the calculator in SEQ mode.**

   To highlight an item in the Mode menu, use the ▶◀▲▼ keys to place the cursor on the item, and then press ENTER. Highlight **SEQ** in the fifth line to put the calculator in Sequence mode. See the first screen in Figure 15-1.

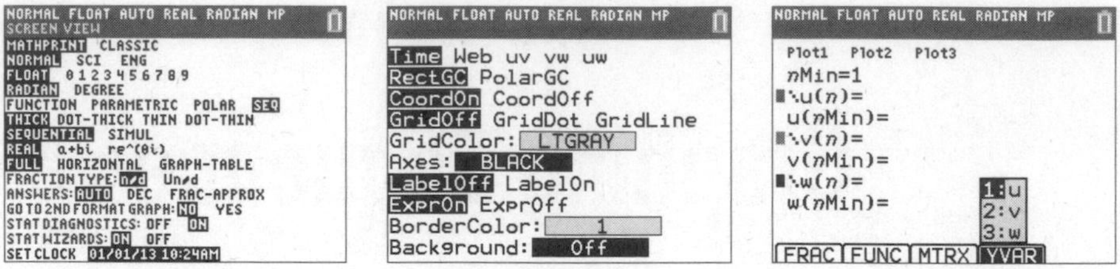

Figure 15-1:
Changing
the mode to
Sequence
and Time
plot.

SEQ mode	Format menu	Press [2nd] [TRACE]

2. **Press [2nd][ZOOM] to access your Format menu.**

The top line in the Format menu is where you choose the type of sequence plot you want to graph. Here are the five choices:

- **Time:** Time plots are discreet graphs (meaning the points are unconnected) of sequences. This is the most popular type of plot. The independent variable, n, will be on the x-axis and $u(n)$, $v(n)$, or $w(n)$ will be on the y-axis. I chose Time in the second screen in Figure 15-1.

- **Web:** Web plots are used to study recursive sequences that converge, diverge, or oscillate. A web plot enables you to see the behavior of the sequence over the long-term. The x-axis is one of the recursive variables, $u(n–1)$, $v(n–1)$, or $w(n–1)$. The y-axis is $u(n)$, $v(n)$, or $w(n)$.

- **uv, vw, and uw:** Phase plots are used to show relationships between two sequences. **uv** has an x-axis of $u(n)$ and y-axis $v(n)$. **vw** has an x-axis of $v(n)$ and y-axis $w(n)$. **uw** has an x-axis of $u(n)$ and y-axis $w(n)$.

On a TI-84 Plus C, press [ALPHA][TRACE] to enter the variables u, v, and w as shown in the third screen in Figure 15-1. If you are using a TI-84 Plus, press [2nd] followed by the [7],[8], or [9] key.

Selecting the Color

All the functionality that you are used to having in Function mode is also available in Sequence mode. If you would like to customize your graph by changing the color or the line style, follow these steps:

1. **Press [Y=] to access the Y= editor.**

2. **Press [◄] to navigate your cursor to the left of $u(n)$.**

See the first screen in Figure 15-2. Graphing a sequence on your calculator requires using up to three lines in the Y= editor.

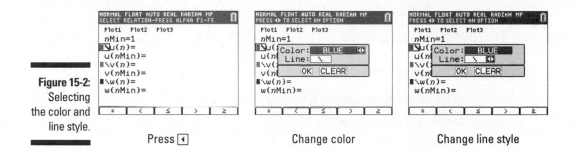

Figure 15-2:
Selecting
the color and
line style.

Press ◄   Change color   Change line style

3. **Press** ENTER **and use the** ▶◄ **keys to change the color using the spinner menu.**

   See the second screen in Figure 15-2.

4. **Press** ENTER **and use the** ▶◄ **keys to change the line style using the spinner menu.**

   I recommend keeping the line style at Dotted–Thick (ᐧ) or changing to Dotted–Thin (ᐧ·). Time plots should be discreet graphs.

5. **Press** ENTER **twice to make the changes effective.**

Entering Sequences

If you are an observant person, then you may have noticed the Y= editor looks quite different than you may be used to. It seems like Y_1 has been replaced with three equations nMin, $u(n)$, and $u(n$Min). Remember, the x and y variables are now defined in terms of a new parameter, n. When you press X,T,Θ,n in Sequence mode, an n appears instead of an x.

Here are a few terms that you need to begin sequence graphing:

- ✔ **n:** The independent variable.

- ✔ **nMin:** Where n starts counting (usually at 1).

- ✔ **$u(n)$:** This is the function that generates the sequence.

- ✔ **$u(n$Min):** The value of the initial term. This is not necessary to include unless you are dealing with recursive sequences. If you need to include more than one term, use {} and enter the terms as a list.

- ✔ **$u(n-1)$:** Previous term, used in *recursive* sequences.

- ✔ **$u(n-2)$:** Term before the previous term, used in *recursive* sequences.

For this exercise, I use this arithmetic sequence: 2, 5, 8, . . .

Explicit formula for an arithmetic sequence: $a_n = a_1 + (n-1)d$

The explicit formula for this sequence is $a_n = 2 + (n-1)3$. Once you have found the explicit formula for a particular sequence, you can enter the sequence in your calculator by following these steps:

1. **Press** $\boxed{\text{Y=}}$ **to access the Y= editor.**

2. **Enter a value for** *n***Min.**

   nMin is the value where n starts counting; I usually enter **1**. See the first screen in Figure 15-3.

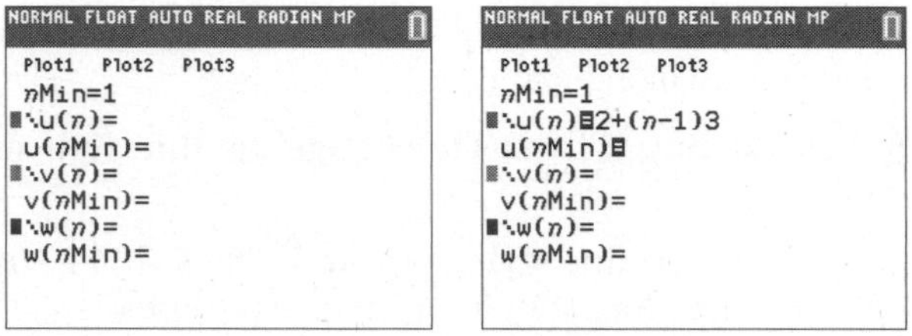

Figure 15-3:
Entering
sequences.

Enter nMin   Enter u(n)

3. **Press** $\boxed{\text{ENTER}}$ **or press** $\boxed{\vee}$ **to navigate to the next line.**

4. **Enter the explicit formula for** *u(n)***.**

   See the second screen in Figure 15-3. Don't forget to press $\boxed{\text{X,T,}\Theta\text{,}n}$ for n.

5. **Press** $\boxed{\text{ENTER}}$ **or press** $\boxed{\vee}$ **to navigate to the next line.**

6. **Enter** *u(n***Min) if necessary.**

   Only enter a value for the initial term if it is a *recursive* sequence.

Setting the Window

It's difficult to get a nice viewing window for your sequence on your first attempt. Understanding the basics of sequence notation should help. Here are the variables used in the window of a sequence graph:

✔ *n***Min:** Where n starts counting. I usually enter **1** (which is the default value).

✔ *n***Max:** Where n stops counting. I recommend choosing a value for nMax that is as large as you might need.

✔ **PlotStart:** First term number to be plotted; I usually enter **1** (which is the default value).

✔ **PlotStep:** Increment for the *n* value, used only in graphing. I usually enter **1** (which is the default value).

Remember, use the ▶◀▲▼ keys to navigate the Window menu. Here are the steps to set your graphing window:

1. **Press** WINDOW **to access the Window editor.**

   See the first screen in Figure 15-4.

   In most cases, it is best to set *n*Min, PlotStart, and PlotStep to **1**. I am going to leave those values at their default value of **1**.

```
NORMAL FLOAT AUTO REAL RADIAN MP     ▯
WINDOW
 nMin=1
 nMax=10
 PlotStart=1
 PlotStep=1
 Xmin=-10
 Xmax=10
 Xscl=1
 Ymin=-10
↓Ymax=10
```

```
NORMAL FLOAT AUTO REAL RADIAN MP     ▯
WINDOW
 nMin=1
 nMax=100
 PlotStart=1
 PlotStep=1
 Xmin=0
 Xmax=10
 Xscl=1
 Ymin=0
↓Ymax=25
```

Figure 15-4: Setting the window.

   Window editor   Enter values

2. **Enter *n*Max.**

   I recommend choosing a value of *n*Max that is as large as you might need; I usually enter **100**. See the second screen in Figure 15-4.

3. **Enter the Xmin.**

   I usually enter **Xmin = 0** for an aesthetically pleasing graph.

4. **Enter the Xmax.**

   How many terms do you want to graph? Assuming the PlotStep and *n*Min are **1**, enter the number of terms you want to graph.

5. **Enter the Ymin.**

   Enter a value a little smaller than the smallest *y*-value in the sequence. Sometimes you have to guess this value; in this sequence, the smallest *y*-value is 2, so I entered **0**.

6. **Enter the Ymax.**

   Enter a value a little larger than the largest *y*-value in the sequence. Sometimes you have to guess this value; in this sequence, I guessed the largest *y*-value would be around 25, as shown in the second screen in Figure 15-4.

If the values of Xmax or Ymax are large, you might want to adjust the Xscl or Yscl values accordingly.

Graphing Sequences

Okay. You are ready to graph your sequence. If your sequence doesn't display as nicely as you had hoped, press WINDOW and adjust the variables accordingly.

Press GRAPH. See the graph in Figure 15-5. It looks like the Xmax needs to be larger so that all the terms in the sequence will graph.

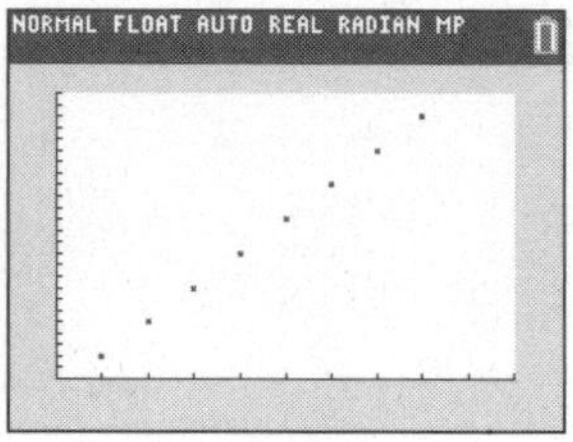

Figure 15-5:
Graphing
sequences.

Using Trace to Evaluate a Sequence Equation

You are going to love using the Trace feature to evaluate sequences. Remember, you are not tracing *x*-values as you do in Function mode. Follow these steps to evaluate a sequence at specific *n* values:

1. Press TRACE.

See the first screen in Figure 15-6. Your trace starts at the smallest *n* value on the graph. The values of X and Y are displayed in the border at the bottom of your graph screen.

The TI-84 Plus C displays functions and information in the border of the graph screen. The TI-84 Plus displays similar information directly on the graph screen.

2. Press ▶ **to find the next term in the sequence.**

Notice, your cursor jumps from one point to the next.

3. Enter a specific *n* value.

After pressing TRACE, entering a number opens up an entry line in the border at the bottom of your graph screen.

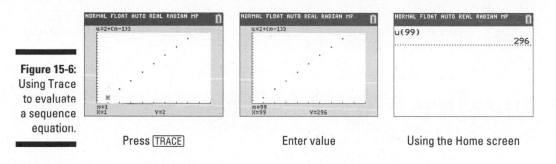

Figure 15-6:
Using Trace
to evaluate
a sequence
equation.

Press [TRACE]   Enter value   Using the Home screen

4. Press [ENTER].

I found something completely unexpected! While using Trace, you can
enter a specific n value that is larger than the Xmax. As long as you
enter a value smaller than nMax, you won't get an error message! See the
second screen in Figure 15-6.

You don't need a graph to evaluate a sequence. Instead, use your calculator by pressing [2nd][MODE]. Enter u by pressing [ALPHA][TRACE][ENTER], then
press [(] and the n-value you would like to evaluate. See the third screen
in Figure 15-6.

Viewing the Table of a Sequence Graph

Sometimes, it is helpful to take a numeric view of the sequence. Press
[2nd][GRAPH] to view the table as shown in the first screen in Figure 15-7.

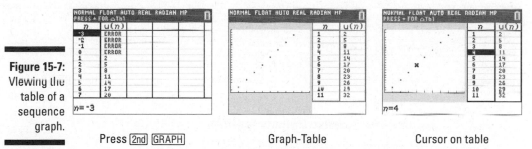

Figure 15-7:
Viewing the
table of a
sequence
graph.

Press [2nd] [GRAPH]   Graph-Table   Cursor on table

Negative values of n will display an error in the table. Press [2nd][WINDOW] and
change the TblStart to 1 to avoid displaying errors.

Another option is to show a split screen with a graph and a table. Press [MODE],
use the [▶][◀][▲][▼] keys to highlight GRAPH-TABLE, and press [ENTER]. Press
[GRAPH] to see the split screen as shown in the second screen in Figure 15-7.

TIP

If you press any key used in graphing functions, such as GRAPH or TRACE, the cursor becomes active on the graph side of the screen. To return the cursor to the table, press 2nd GRAPH. See the third screen in Figure 15-7.

Graphing a Recursive Sequence

In order to contrast explicit and recursive sequences, I am going to use the same arithmetic sequence: 2, 5, 8, . . .

Recursive formula for an arithmetic sequence: $a_n = a_{n-1} + d$

The recursive formula for this sequence is $\boldsymbol{a_n = a_{n-1} + 3}$, where $a_1 = 2$. In this formula, a_{n-1} represents the previous term. In Sequence mode on the calculator, the previous term is $u(n-1)$. Follow these steps to enter a recursive sequence in your calculator:

1. **Press** Y= **to access the Y= editor.**

2. **Enter a value for *n*Min.**

   *n*Min is the value where *n* starts counting. I usually enter **1**.

   REMEMBER

   Press TRACE to enter the variables, *u*, *v*, and *w*, as shown in the first screen in Figure 15-8.

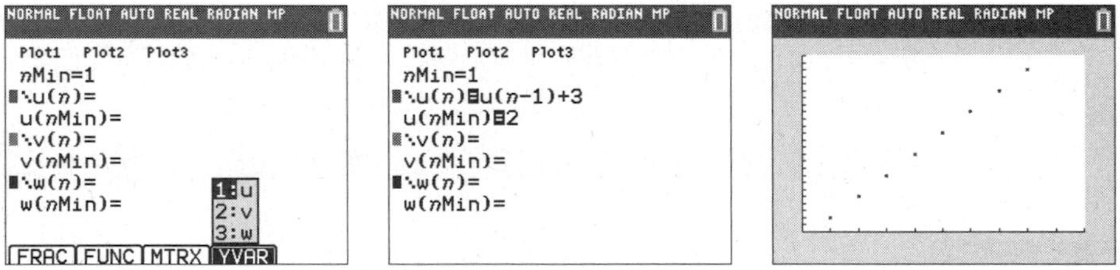

Figure 15-8:
Graphing a
recursive
sequence.

NORMAL FLOAT AUTO REAL RADIAN MP	NORMAL FLOAT AUTO REAL RADIAN MP	NORMAL FLOAT AUTO REAL RADIAN MP
Plot1 Plot2 Plot3	Plot1 Plot2 Plot3	
*n*Min=1	*n*Min=1	
■\u(*n*)=	■\u(*n*)⊟u(*n*−1)+3	
u(*n*Min)=	u(*n*Min)⊟2	
■\v(*n*)=	■\v(*n*)=	
v(*n*Min)=	v(*n*Min)=	
■\w(*n*)= **1:u**	■\w(*n*)=	
w(*n*Min)= 2:v	w(*n*Min)=	
3:w		
FRAC FUNC MTRX **YVAR**		

Access u, v, and w   Window editor   Press GRAPH

3. **Enter the recursive formula for *u*(*n*).**

   Don't forget to press X,T,Θ,*n* for *n*.

4. **Enter *u*(*n*Min).**

   Enter the initial term as shown in the second screen in Figure 15-8.

5. **Press** GRAPH.

   The graph of this sequence looks exactly like the graph of the explicit function as shown in the third screen in Figure 15-8.

Graphing the Fibonacci sequence

Want to try something more difficult? The Fibonacci sequence is one of the most famous sequences in mathematics: 1, 1, 2, 3, 5, 8, . . .

The recursive formula for the Fibonacci sequence is: $a_n = a_{n-1} + a_{n-2}$. Remember, $u(n-2)$ means the term before the previous term in calculator lingo.

1. **Press [Y=] to access the Y= editor.**

2. **Enter a value for *n*Min.**

   *n*Min is the value where *n* starts counting. I usually enter **1**.

   Press [2nd][TRACE] to enter the variables *u*, *v*, and *w*.

3. **Enter the recursive formula for *u*(*n*).**

4. **Enter *u*(*n*Min).**

   Enter the first two terms in the sequence as a list! Press [2nd][(] to use brackets for your list. See the first screen in Figure 15-9.

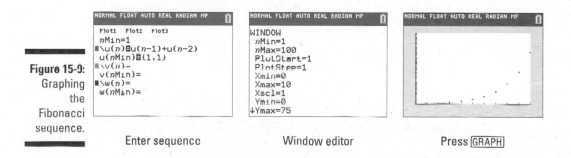

Figure 15-9: Graphing the Fibonacci sequence.

Enter sequence   Window editor   Press [GRAPH]

5. **Press [WINDOW] and adjust the variables.**

   I changed the value of Ymax. See the second screen in Figure 15-9.

6. **Press [GRAPH].**

   See the third screen in Figure 15-9.

   Did you notice all the tick marks on the *y*-axis in the third screen in Figure 15-9? It would have been a good idea to change the Yscl to 5 or 10. Press [WINDOW] to access the Yscl variable.

Graphing a Web Plot

Web plots are a really interesting representation of the behavior of a recursive sequence. Web plots actually look a little like spider webs, but I think I am getting ahead of myself. First, a few ground rules when working with web plots:

🖊 **Sequence must be recursive:** The sequence is required to have one level of recursion $u(n-1)$; using $u(n-2)$ is not allowed.

🖊 **Sequence cannot reference n directly.**

🖊 **Sequence cannot reference any defined sequence except itself.**

I will use the recursive sequence, $a_n = -\frac{1}{2}(a_{n-1})$, where $a_1 = 8$.

Here are the steps to enter and graph a web plot in your calculator:

1. **Press 2nd ZOOM to access the Format menu.**

2. **Use the ▶◀▲▼ keys to navigate your cursor to Web in the top row and press ENTER.**

3. **Press Y= to access the Y= editor.**

4. **Enter the recursive formula for $u(n)$.**

   Use $u(n-1)$ for the previous term.

5. **Enter $u(n\text{Min})$.**

   Enter the initial term in the sequence as shown in the first screen in Figure 15-10.

6. **Press ZOOM 6.**

   If you aren't sure what the graph of the web plot will look like, a ZStandard graphing window is a good place to start, as shown in the second screen in Figure 15-10.

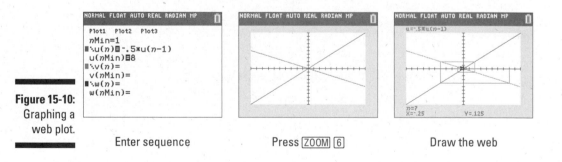

Figure 15-10:
Graphing a
web plot.

Enter sequence   Press ZOOM 6   Draw the web

The line graphed at $y = x$ is not part of the web plot; it is called the reference line.

7. **Press TRACE and press ▶ repeatedly to draw the web.**

   See the third screen in Figure 15-10. The web plot helps you see that this sequence converges.

   To clear a webplot, press 2nd PRGM ENTER.

Graphing a Phase Plot

If you want to see how one sequence affects another sequence, you can use a phase plot to represent the data.

I will enter two sequences that represent the population of rabbits and foxes. Of course, the population of predators is related to the population of its prey.

In the sequence for rabbits: $u(n) = u(n–1)*(1.05–0.001*v(n–1)$, initial condition = 200, $0 < n < 500$.

In the sequence for foxes, $v(n)=v(n–1)*(0.97+0.0002*u(n–1)$, initial condition = 50, $0 < n < 500$.

Here are the steps to enter and graph a phase plot in your calculator:

1. **Press** 2nd ZOOM **to access the Format menu.**

2. **Use the** ▶◀▲▼ **keys to navigate your cursor to "uv" in the top row and press** ENTER.

3. **Press** Y= **to access the Y= editor.**

4. **Enter the recursive formula for *u*(*n*) and *v*(*n*).**

5. **Enter *u*(*n*Min) and *v*(*n*Min).**

   Enter the initial term in each sequence as shown in the first screen in Figure 15-11.

6. **Press** WINDOW **and adjust the variables.**

   Make sure your *n*Max = 500, as shown in the second screen in Figure 15-11.

7. **Press** GRAPH.

   See the third screen in Figure 15-11.

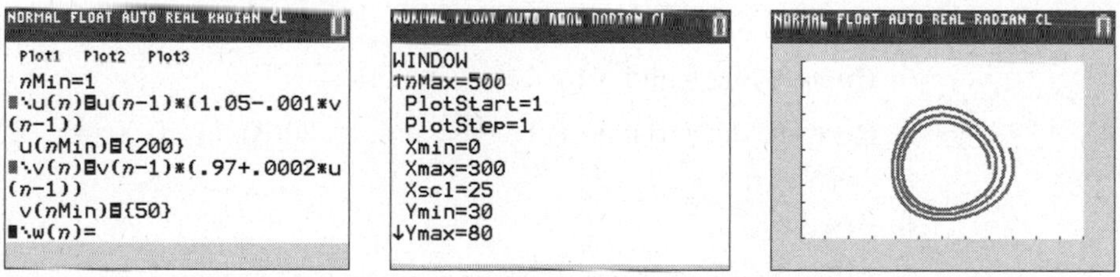

Figure 15-11:
Graphing a
phase plot.

Enter sequences

Press WINDOW

Press GRAPH

Graphing Partial Sums of an Infinite Series

An infinite geometric series has a sum if $-1 < r < 1$. I wonder if the graph of the partial sums would be interesting? The partial sums should approach a number, which means the graph should have an asymptote. Sounds interesting to me!

I'll use the infinite geometric sequence: $S = \Sigma(\frac{1}{2^n})$ as n goes from 0 to infinity.

Here are the steps to graph the partial sums of an infinite series:

1. **Press** 2nd ZOOM, **highlight TIME, and press** ENTER.

2. **Press** Y= **to access the Y= editor.**

3. **Enter a value for nMin.**

   nMin is the value where n starts counting; I usually enter **1**.

4. **Press** ALPHA WINDOW 2 **to use the summation template to enter $u(n)$.**

   See the first screen in Figure 15-12. Press X,T,Θ,n for n, and enter the infinite series pictured in the second screen in Figure 15-12.

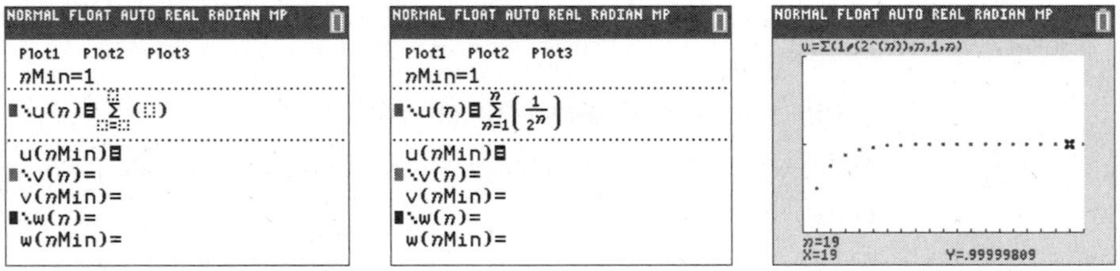

Figure 15-12: Graphing partial sums of an infinite series.

Summation template   Enter series   Press GRAPH

5. **Press** WINDOW **and adjust the variables.**

   Here are the variables I changed: nMax = 20, Xmin = 0, Xmax = 20, Ymin = 0, Ymax = 2.

6. **Press** GRAPH.

7. **Press** TRACE **and use** ▶ **to find the partial sums.**

   See the third screen in Figure 15-12.

Part IV
Working with Probability and Statistics

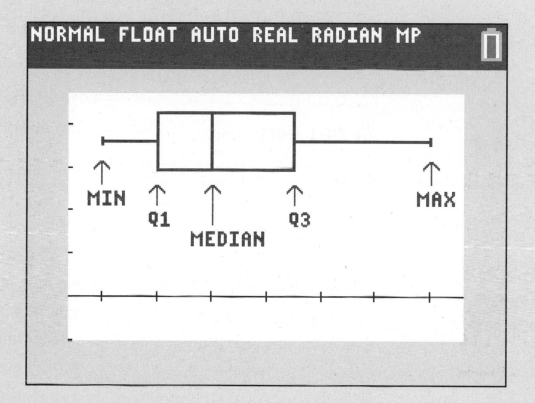

In this part . . .

✔ Get a look at calculating factorials, permutations, combinations, and even generating random numbers.

✔ See how to enter and sort data in a list and use formulas to generate a data list.

✔ Find out how to use regression to find the curve of best fit and display the data as a histogram or box plot.

Chapter 16

Probability

Do you need to calculate the number of ways you can arrange six people at a table or the number of ways you can select four people from a group of six people? Are you learning about factorials or the Binomial theorem in math class? Or do you just need an unbiased way of selecting people at random? If so, this is the chapter for you.

Evaluating Factorials

Did you know you can type an exclamation point on your calculator? Mathematically, the exclamation point is called a *factorial*. Usually students learn about factorials in pre-algebra and then forget what they are by the time they need to use factorials to solve tough probability problems. Here is a quick refresher on factorials.

4! = 4*3*2*1 and 7! = 7*6*5*4*3*2*1. See Figure 16-1. If you haven't done so already, press 2nd MODE to get to the Home screen. All the calculations in the chapter use the Home screen. Follow these steps to type a factorial in your calculator:

[handwritten: MATH →→ → PROB choose 4 enter]

1. **Enter the number you would like to take the factorial of.**

2. **Press** MATH ◀ ◀ **to access the Math Probability menu, and press** 4 **to choose the factorial symbol (it looks like an exclamation point.)**

There are more MATH submenus available on the TI-84 Plus C, if you use the TI-84 Plus, pay attention to the name of the submenu and use the ◀ ▶ keys to navigate to the correct one.

3. **Press** ENTER **to evaluate the factorial.**

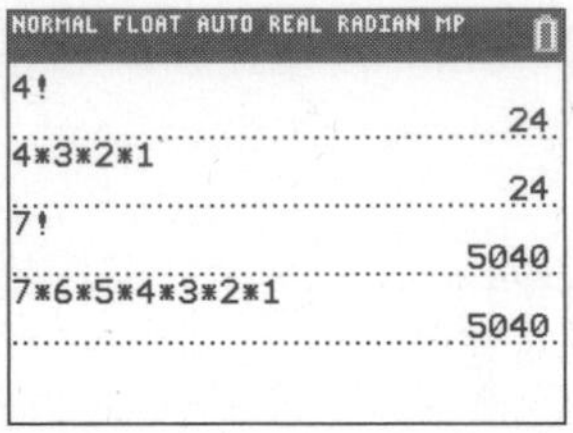

Figure 16-1:
Evaluating
factorials.

Handwritten annotations: "nCr", "out of total # of items", "nPr", "how many #s to arrange", "indep. of order"

Permutations and Combinations

Handwritten annotation: "order is important"

A *permutation,* denoted by **nPr**, answers the question: "From a set of **n** different items, how many ways can you select *and* order (arrange) **r** of these items?" One thing to keep in mind is that order is important when working with permutations. Permutation questions may ask questions like, "In how many ways could ten runners end up on the Olympic medal stand (Gold, Silver, or Bronze)?" Is order important? Yes; use **nPr** with **n = 10** and **r = 3**).The formula for a permutation is: **nPr = (n!)/(n-r)!**

A *combination,* denoted by **nCr**, answers the question: "From a set of **n** different items, how many ways can you select (independent or order) **r** of these items?" Order is not important with combinations. Combination questions may look like, "A subcommittee made up of 4 people must be selected from a group of 20." Is order important? No; the five committee positions are equally powerful. Use **nCr** with **n = 20** and **r = 4**. The formula for a combination is: **nCr = (n!)/(r!(n-r)!)**.

Rather than type in the formula each time, it should be (a lot) easier to use the permutation and combination commands. To evaluate a permutation or combination, follow these steps:

Handwritten annotations: "[MATH] → Prob", "choose 2 for Perm", "choose 3 for Combo", "OR", "TIP", "[alpha] [window]", "choose 7 Perm", "choose 8 Combo"

1. **There are two ways to access the nPr and nCr templates: Press**
   MATH ◄ ◄ **to access the Math PROB menu or press** ALPHA WINDOW **to access the shortcut menu.**

   On the TI-84 Plus, press MATH ◄ to access the probability menu where you will find the permutations and combinations commands. Using the TI-84 Plus, you must enter *n*, insert the command, and then enter *r*.

   See the PROB menu in the first screen in Figure 16-2. See the shortcut menu in the second screen in Figure 16-2. Press the number on the menu that corresponds to the template you want to insert.

2. **In the first blank, enter n, the total number of items in the set.**

   Alternatively, you could enter **n** first and then insert the template.

3. **Press ▶ to navigate your cursor to the second blank in the template.**

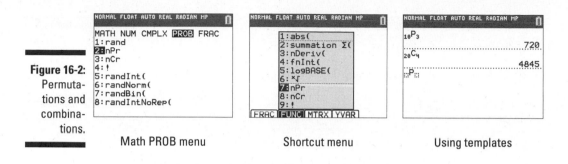

Figure 16-2:
Permutations and combinations.

Math PROB menu   Shortcut menu   Using templates

4. **Enter r, the number of items selected from the set, and press** ENTER **to display the result.**

See the last screen in Figure 16-2. Notice the blank **nPr** template in the last line of the last screen in Figure 16-2.

Using the Binomial theorem

In math class, you may be asked to expand binomials. This isn't too bad if the binomial is $(2x+1)^2 = (2x+1)(2x+1) = 4x^2 + 4x + 1$. That's easy. What if you were asked to find the fourth term in the binomial expansion of $(2x+1)^7$? Now that is more difficult.

The general term of a binomial expansion of **(a+b)ⁿ** is given by the formula: **(nCr)(a)ⁿ⁻ʳ(b)ʳ**. To find the fourth term of $(2x+1)^7$, you need to identify the variables in the problem:

- ✔ **a:** First term in the binomial, **a = 2x**.

- ✔ **b:** Second term in the binomial, **b = 1**.

- ✔ **n:** Power of the binomial, **n = 7**.

- ✔ **r:** Number of the term, but **r** starts counting at **0**. This is the tricky variable to figure out. My students think of this as one less than the number of the term you want to find. Since you want the fourth term, **r = 3**.

Plugging into your formula: **(nCr)(a)ⁿ⁻ʳ(b)ʳ** = **(7C3) (2x)⁷⁻³(1)³**.

Evaluate **(7C3)** in your calculator:

1. **Press** ALPHA WINDOW **to access the shortcut menu.**

See the first screen in Figure 16-3.

2. **Press** 8 **to choose the nCr template.**

See the first screen in Figure 16-3.

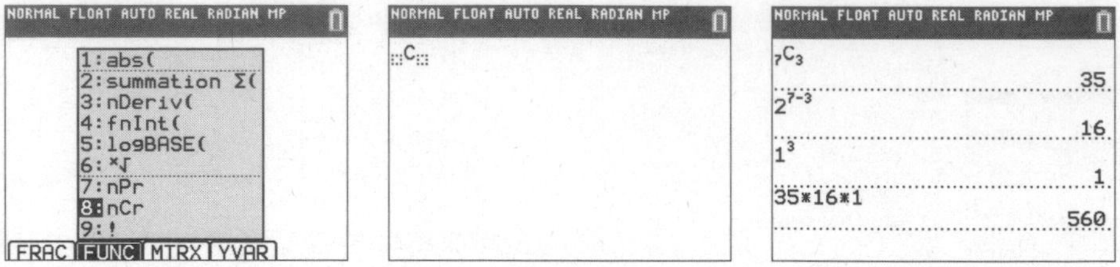

NORMAL FLOAT AUTO REAL RADIAN MP

```
1:abs(
2:summation Σ(
3:nDeriv(
4:fnInt(
5:logBASE(
6:×√
7:nPr
8:nCr
9:!
```
FRAC FUNC MTRX YVAR

NORMAL FLOAT AUTO REAL RADIAN MP

$_\square C_\square$

NORMAL FLOAT AUTO REAL RADIAN MP

$_7C_3$

$\qquad\qquad\qquad\qquad 35.$

2^{7-3}

$\qquad\qquad\qquad\qquad 16.$

1^3

$\qquad\qquad\qquad\qquad 1.$

$35\times16\times1$

$\qquad\qquad\qquad\qquad 560.$

Figure 16-3:
Using the
Binomial
theorem.

Press ALPHA WINDOW　　　　nCr template　　　　Calculations

On the TI-84 Plus, press MATH◄ to access the probability menu where you will find the permutations and combinations commands. Using the TI-84 Plus, you must enter n, insert the command, and then enter r.

3. Enter n in the first blank and r in the second blank.

Alternatively, you could enter **n** first and then insert the template.

4. Press ENTER to evaluate the combination.

5. Use your calculator to evaluate the other numbers in the formula, then multiply them all together to get the value of the coefficient of the fourth term.

See the last screen in Figure 16-3. The fourth term of the expansion of $(2x+1)^7$ is $560x^4$.

Generating Random Numbers

Your calculator has a massive amount of digits arranged in a list, called a random number table, that it uses to generate random numbers. Some math textbooks have a random number table in the appendix. I guess you don't really have to know that, but it helps you understand how a random number can be "seeded," as explained at the end of this chapter. In this chapter, I save the best for last.

Generating random integers

To generate random integer, follow these steps:

1. Press MATH◄◄5 to activate the randInt Wizard from the Math PROB menu.

If you are using a TI-84 Plus, there is no wizard for the randint command. To use the command, you must know the syntax: randint(lower, upper, [number of elements]).

A wizard makes entering information easy. A wizard arranges the data you enter so that it fits the syntax of the command. See the wizard in the first screen in Figure 16-4.

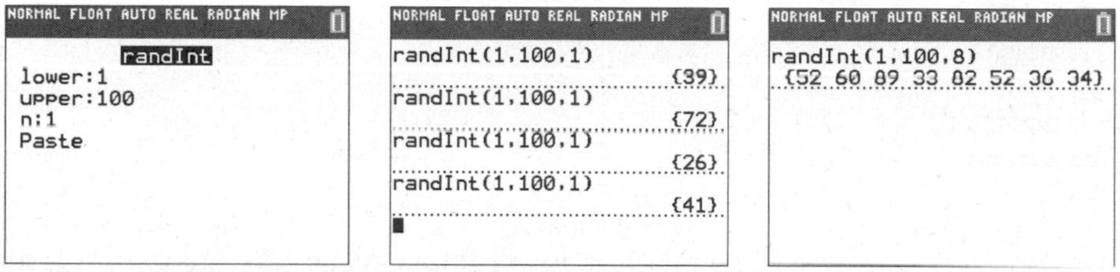

Figure 16-4:
Generating random integers.

NORMAL FLOAT AUTO REAL RADIAN MP

randInt
lower:1
upper:100
n:1
Paste

randInt Wizard

NORMAL FLOAT AUTO REAL RADIAN MP

randInt(1,100,1)
{39}
randInt(1,100,1)
{72}
randInt(1,100,1)
{26}
randInt(1,100,1)
{41}

Press ENTER

NORMAL FLOAT AUTO REAL RADIAN MP

randInt(1,100,8)
{52 60 89 33 82 52 36 34}

Generating a list

2. **Enter the lower limit and upper limit you want your random number to be.**

   I want a random number from 1 to 100. Press ENTER or ▾ to navigate to the next line in the wizard.

3. **Enter n, for how many random numbers you want to generate.**

   Press ENTER repeatedly to generate more random numbers as illustrated in the second screen in Figure 16-4.

4. **If you want to generate a list of random integers, change the value of** n.

   See the third screen in Figure 16-4. I changed **n** to **8**.

Generating random integers with no repetition

Did you notice the integer, 52, was selected twice in the third screen in Figure 16-4? If you are generating a list of random integers, you can easily avoid repeats by using a different command. Here are the steps:

1. **Press** MATH ◀ ◀ 8 **to activate the randIntNoRep Wizard from the Math PROB menu.**

   See the first screen in Figure 16-5.

   On the TI-84 Plus, the randIntNoRep command does not have a wizard to help you. The syntax for the command is randIntNoRep(lower,upper). Unlike the TI-84 Plus C, you don't have the option of adjusting the number of terms.

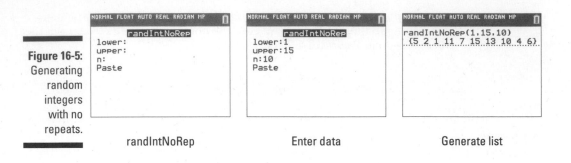

Figure 16-5: Generating random integers with no repeats.

randIntNoRep   Enter data   Generate list

2. **Enter the upper and lower limits and** n **(the number of terms).**

   See the second screen in Figure 16-5.

3. **Press** ENTER **until your numbers have been generated.**

   This is illustrated in the third screen in Figure 16-5.

Generating random decimals

It is easy to generate random decimal numbers that are strictly between 0 and 1. Press MATH ◄ ◄ ENTER to select the **rand** command from the Math Probability menu. Then repeatedly press ENTER to generate the random numbers. The first screen in Figure 16-6 illustrates this process.

To generate random numbers between 0 and 100, use the rand command in an expression: **100*rand**. See the second screen in Figure 16-6.

Figure 16-6: Generating random numbers (between 0 and 1).

Decimals   From 0 to 100

Seeding the random number generator

Earlier in the chapter, I mention your calculator generates random numbers from a massive list of digits arranged in a list. Here is the cool part. You can pick where in the list you want your calculator to start generating random numbers. It is called seeding your random number. In a class, I can have all my students seed their calculators using their phone numbers. Each student's calculator generates different random numbers based on the seed he or she selects.

Let's get a little creative with the number you select to seed your calculator. This should be fun! Let's figure out how many days you have been alive. Your calculator has a command that can figure that out for you! Follow these steps:

1. **Press** 2nd 0 **to access the Catalog.**

   Notice, your calculator is in Alpha mode, indicated by the blinking A in the cursor.

2. **Press** x⁻¹ ▼ ENTER **to insert the** dbd(**function.**

   dbd stands for *days between dates.*

3. **Enter your birth date as a number in this form:** MM.DDYY.

   Dates must be between the years 1950 and 2049. I entered **June 1, 1968**, with the number: **06.0168**.

4. **Press** , .

5. **Enter today's day as a number in this form:** MM.DDYY.

   I entered **January 2, 2013**, with the number: **01.0213**.

6. **Enter** ENTER **to find out how many days you have been alive.**

   See the first screen in Figure 16-7. Wow! 16,286 days sounds old!

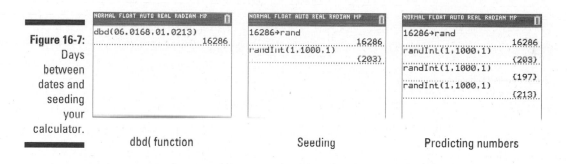

Figure 16-7:
Days
between
dates and
seeding
your
calculator.

dbd(function   Seeding   Predicting numbers

Here are the steps to seed your calculator:

1. **Enter the number you are using to seed your calculator.**

   I entered **16286**. Of course, you could use any real number to seed your calculator.

2. Press STO▸.

3. Press MATH ◄ ◄ ENTER to insert the rand command.

4. Press ENTER to seed your calculator.

   See the first line in the second screen in Figure 16-7.

5. **Try it out! Use** randInt(**to generate a random number.**

   See the last line in the second screen in Figure 16-7.

Want to impress your friends? Seed your calculator with results you know in advance. For example, if you secretly seed your calculator with the number **16286**, then the next three random numbers (from 0 to 1,000) that will be generated will be 203, 197, and 213, as shown in the third screen in Figure 16-7. Just don't share your secret!

Chapter 17

Dealing with Statistical Data

· ·

· ·

*T*he calculator has many features that provide information about the data entered in the calculator. It can graph data as a scatter plot, histogram, or box plot. The calculator can calculate the median and quartiles. It can even find a regression model (curve fitting) for your data. It can do this and much, much more. This chapter explains how to enter your data in the calculator; Chapter 18 shows you how to use the calculator to analyze that data.

Entering Data

What you use to enter statistical data into the calculator is the Stat List editor — a relatively large spreadsheet that can accommodate up to 20 columns (data lists). And each data list (column) can handle a maximum of 999 entries. Pictures of the Stat List editor appear in Figure 17-1.

To use stat lists to enter your data into the calculator, follow these steps:

1. **Press** STAT **to access the Stat EDIT menu.**

   See the first screen in Figure 17-1.

2. **Press** 5 ENTER **to execute the SetUpEditor command.**

   The SetUpEditor command clears all data lists (columns) from the Stat List editor and replaces them with the six default lists L_1 through L_6. Any lists that are cleared from the editor by this command are still in the memory of the calculator; they just don't appear in the Stat List editor.

Figure 17-1:
The Stat List editor.

Stat EDIT menu   Empty lists   Lists with data

3. **Press** ⌊STAT⌋ ⌊ENTER⌋ **to enter the Stat List editor.**

   If no one has ever used the Stat List editor in your calculator, then the Stat List editor looks like the second screen in Figure 17-1. If the Stat List editor has been used before, then some of the default lists L_1 through L_6 may contain data, as in the third screen in Figure 17-1.

4. **If necessary, clear lists L_1 through L_6.**

   When you clear a data list, the list's contents (and not its name) will be erased, leaving an empty data list in the calculator's memory. To clear the contents of a data list in the Stat List editor, use the ⌊▶⌋⌊◀⌋⌊▲⌋⌊▼⌋ keys to place the cursor on the name of a list appearing in a column heading, as shown in the first screen in Figure 17-2. To clear the list, press ⌊CLEAR⌋ and don't panic when nothing seems to happen! Now, press ⌊ENTER⌋ or ⌊▼⌋ to see the list contents disappear, as shown in the second screen in Figure 17-2.

   An alternative method of clearing a list is to press ⌊STAT⌋⌊4⌋ on the Home screen to insert the ClearList command. Then press ⌊2nd⌋⌊STAT⌋, use the ⌊▲⌋⌊▼⌋ keys to choose the list you want to clear, and press ⌊ENTER⌋. To clear multiple lists at one time, place commas between the list names, as shown in the third screen in Figure 17-2.

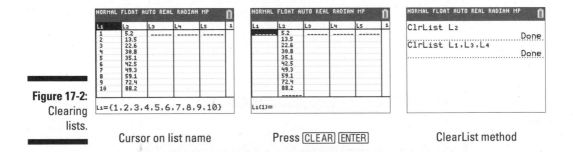

Figure 17-2:
Clearing lists.

Cursor on list name   Press ⌊CLEAR⌋ ⌊ENTER⌋   ClearList method

5. **Enter your data. Press ENTER after each entry.**

Use the ▸◂▴▾ keys to place the cursor in the column where you want to make an entry. Use the keypad to enter your number and press ENTER when you're finished. A column (list) can accommodate up to 999 entries.

Deleting and Editing Data

Sooner or later, you'll have to remove or modify the data that you've placed in a data list. The following descriptions show you how to do so:

✔ **Deleting a data list from the memory of the calculator:**

You can permanently remove a data list from the memory of the calculator. Press 2nd+2 to enter the Memory Management menu, as shown in the first screen in Figure 17-3. Then press 4 to see the data lists that are stored in memory. Use ▾ to move the indicator to the list you want to delete, as shown in the second screen in Figure 17-3. Press DEL to delete that list. When you're finished deleting lists from memory, press 2nd MODE to exit (quit) the Memory Management menu and return to the Home screen.

Although the calculator does enable you to delete default list names (L_1 through L_6) from memory, in reality, it deletes only the contents of the list and not its name.

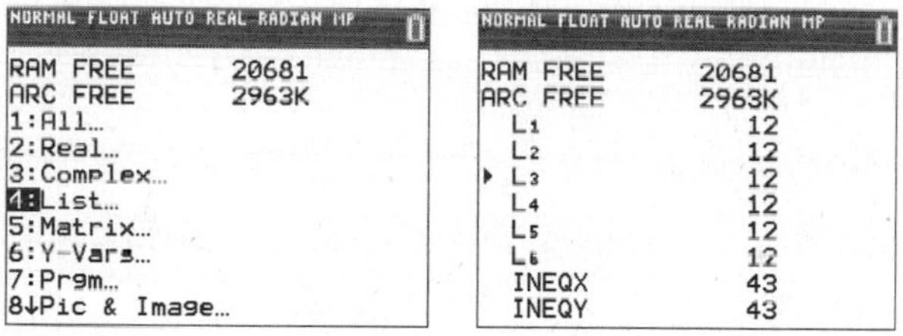

Figure 17-3:
Deleting
lists.

```
NORMAL FLOAT AUTO REAL RADIAN MP
RAM FREE        20681
ARC FREE        2963K
1:All…
2:Real…
3:Complex…
4:List…
5:Matrix…
6:Y-Vars…
7:Prgm…
8↓Pic & Image…
```
Memory Management menu

```
NORMAL FLOAT AUTO REAL RADIAN MP
RAM FREE        20681
ARC FREE        2963K
    L₁          12
    L₂          12
▸   L₃          12
    L₄          12
    L₅          12
    L₆          12
  INEQX         43
  INEQY         43
```
Delete list

✔ **Deleting a column (list) in the Stat List editor:**

To delete a column (list) from the Stat List editor, use the ▸◂▴▾ keys to place the cursor on the name of the list appearing in the column headings, and then press DEL. The list will be removed from the Stat List editor but not from the memory of the calculator. This is a quick and easy method of deleting a list!

✔ **Deleting an entry in a data list:**

To delete an entry from a data list, use the ▷◁▲▽ keys to place the cursor on that entry, and then press DEL to delete the entry from the list.

✔ **Editing an entry in a data list:**

To edit an entry in a data list, use the ▷◁▲▽ keys to place the cursor on that entry, press ENTER, and then edit the entry or key in a new entry. If you key in the new entry, the old entry is automatically erased. To avoid errors, press ENTER or use the ▽▲ keys when you're finished editing or replacing the old entry.

If you delete some lists and want the six default lists back (L_1 through L_6), press STAT 5 ENTER to use the SetUpEditor command.

Inserting Data Lists

You can't rename a list, so if you want to have a nifty name for your list, it is best to insert a list before you start entering your data. To insert a data list in the Stat List editor, follow these steps:

1. **If necessary, press STAT ENTER to enter the Stat List editor.**

2. **Use the ▷◁▲▽ keys to place the cursor on the column heading where you want your list to appear.**

   Your list is created in a new column that will appear to the left of the column highlighted by the cursor (as shown in the first screen in Figure 17-4).

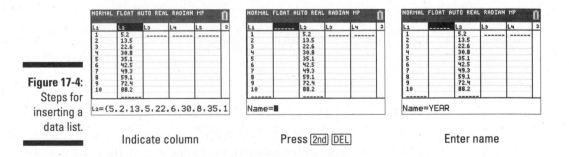

Figure 17-4: Steps for inserting a data list.

Indicate column   Press 2nd DEL   Enter name

3. **Press 2nd DEL to insert the new column.**

   The second screen in Figure 17-4 shows this procedure.

4. **Enter the name of your data list and press** ENTER.

The name you give your data list can consist of one to five characters that must be letters, numbers, or the Greek letter θ. The first character in the name must be a letter or θ.

Press 2nd ALPHA to place the calculator in Alpha-Lock mode. The  after **Name** = indicates that the calculator is in Alpha mode. In this mode, when you press a key, you enter the green letter above the key. To enter a number, exit the mode by pressing ALPHA again, and then enter the number. To enter a letter after entering a number, you must press ALPHA to put the calculator back in Alpha mode (as in the third screen in Figure 17-4). Press ENTER when you're finished entering the name.

Press 2nd ALPHA to put your calculator in Alpha-Lock mode. This enables you to enter letters without pressing ALPHA each time.

If the name you give your data list is the name of a data list stored in memory, then after entering that name and pressing ENTER, the data in the list stored in memory will be automatically entered in the Stat List editor.

After you have named your data list, you can press ⏷ and start entering your data. If the data you want to put in the newly named list is in another column of the Stat List editor — or in a list stored in memory under another name — you can paste that data into your newly named list. (See the section, "Copying and Recalling Data Lists," later in this chapter.)

Using Formulas to Enter Data

Figure 17-5 illustrates how you would place the sequence 10, 20, . . ., 200 in list L_1. The formula used in this example is simply x. The initial and terminal values of x are naturally 10 and 200, respectively. And, as you may guess, x is incremented by 10.

Make sure Stat Wizards are ON in the Mode menu before beginning this section. Press MODE ▲ ▲ ▲ ENTER to turn Stat Wizards ON.

To use a formula to define your data, follow these steps:

1. **If necessary, press** STAT 1 **to enter the Stat List editor.**

2. **Use the** ▶ ◀ ▲ ▼ **keys to place the cursor on the column heading where you want your data to appear, and press** ENTER.

3. **Press** 2nd STAT ▶ 5 **and fill in the Seq Wizard.**

See the first screen in Figure 17-5. The easiest way to create a sequence is to press $\boxed{\text{X,T,Θ,}n}$ for the expression. Fill in the appropriate start and end values. The step is the increment from one term to the next. Wizards just make things easier.

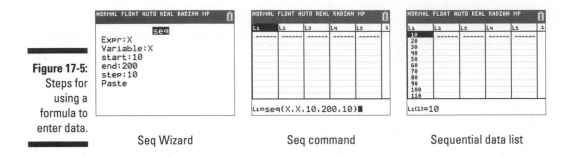

Figure 17-5:
Steps for using a formula to enter data.

                     Seq Wizard   Seq command   Sequential data list

4. **Use the $\boxed{\text{▾}}$ keys to highlight Paste and press $\boxed{\text{ENTER}}$.**

See the second screen in Figure 17-5. Notice how the wizard fills in the syntax of the Seq command for you!

5. **Press $\boxed{\text{ENTER}}$ to create your data in the list.**

This procedure is shown in the third screen in Figure 17-5.

Copying and Recalling Data Lists

Once you have entered your data in a list, you can call the list up again when you want to use or change it.

  ✔ **Copying data from one list to another:**

After you enter data into the Stat List editor, that data is automatically stored in the memory of the calculator under the list name that appears as the column heading for that list. You don't have to take any further steps to ensure that the calculator saves your data. However, if you clear the contents of a data list (as described in the earlier section, "Deleting and Editing Data"), the calculator retains the name of the data list in memory but deletes the contents of that list.

If you enter your data in one of the default lists L_1 through L_6 and would like to save it as a named list, first press $\boxed{\text{2nd}}\boxed{\text{DEL}}$ to insert a data list. Use $\boxed{\text{ALPHA}}$ to enter a name for the list and press $\boxed{\text{ENTER}}$. You will get a result that resembles the first screen in Figure 17-6. Then press $\boxed{\text{2nd}}\boxed{\text{STAT}}$ to access the List NAMES menu and press $\boxed{\text{ENTER}}$ to select L_1, as shown in the second screen in Figure 17-6.

The quickest way to enter **L₁** is to press [2nd][1]. Or, enter **L₂** by pressing [2nd][2]. Notice the tiny blue lettering above keys 1 through 6 on the calculator, indicating their secondary key functions.

Finally, press [ENTER] to insert the data from **L₁** into the newly named data list. The third screen in Figure 17-6 shows this process.

Figure 17-6:
Steps for
copying
data from
one list to
another.

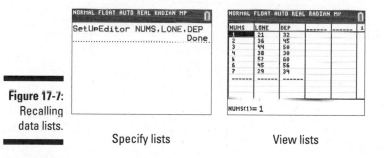

Insert list   Enter L₁   Press [ENTER]

✓ **Recalling data lists:**

You can use the SetUpEditor command to set up the Stat List editor with the data lists you specify. To do this, press [STAT][5] to invoke the SetUpEditor command. Enter the names of the data lists, separated by commas. Then press [ENTER][STAT][1] to see the data lists (as shown in the first two screens in Figure 17-7).

If you're already in the Stat List editor, you can recall a data list by inserting a data list and entering the name of the saved data list.

Figure 17-7:
Recalling
data lists.

Specify lists   View lists

You can save a data list on your PC and recall it at a later date. (Chapter 20 explains how to do this.) You can also transfer a data list from one calculator to another, as described in Chapter 21.

Sorting Data Lists

For this example, I solve a typical standardized test question.

Put the following set of numbers in order from least to greatest:

$$\left\{-\sqrt{(2)},\ -\frac{7}{5},\ -1.25,\ -\frac{3}{2}\right\}$$

To sort a data list, follow these steps:

1. **Press** STAT ENTER **and enter the data in** L_1.

   See the first screen in Figure 17-8. Notice, after entering $-\sqrt{(2)}$ and pressing ENTER, your calculator evaluates the square root and displays its approximate value, **–1.414**.

   If list L_1 is out of view, press STAT 5 ENTER to use the SetUpEditor command. If there is unwanted data in list L_1 use the ► ◄ ▲ ▼ keys to place the cursor on the L_1 list name and press CLEAR ENTER.

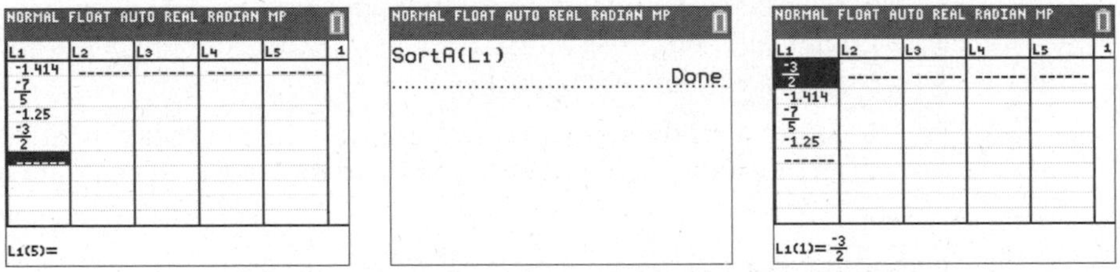

Figure 17-8:
Sorting
data.

| Unsorted data | SortA(command | Sorted data |

2. **Press** STAT.

3. **Press** 2 **to sort the list in ascending order.**

   SortA means sort ascending and SortD means sort descending.

4. **Enter the list name.**

   To sort a default named list such as L_1, press 2nd 1 to enter its name. If you're sorting a list that you named, press 2nd STAT to access the List NAMES menu, use the ▲ ▼ keys to scroll to the list you want, and press ENTER.

   An alternative method to type the name of a list you named is to press 2nd STAT ► ▲ ENTER to insert the letter **L** and then enter the name of the list.

5. **Press** ENTER **to sort list L₁.**

   See the second screen in Figure 17-8.

6. **Press** STAT ENTER **to view list L₁.**

   See the third screen in Figure 17-8. It is easy to see the answer to the question posed is: $\{-3/2, -\sqrt{(2)}, -7/5, -1.25\}$.

Sorting data lists while keeping the rows intact

In most cases, it is a good idea to keep the rows of data intact when sorting. Follow these steps to sort data lists while keeping the rows intact:

1. **Press** STAT ENTER **and enter the data in L₁.**

   See the first screen in Figure 17-9.

Figure 17-9: Sorting data lists while keeping rows intact.

Sorted by YEAR   Sorting multiple lists   Sorted by RATE

2. **Press** STAT.

3. **Press** 2 **or** 3 **to sort the list in ascending or descending order, respectively.**

4. **Enter the list name that you want to sort on.**

   To sort list such as **L₂**, press 2nd 2. If you're sorting a list that you named, press 2nd STAT to access the List NAMES menu, use the ▲▼ keys to scroll to the list you want, and press ENTER.

5. **Press** , **between the data lists you want to sort concurrently.**

6. **Enter the other list name that you want to sort concurrently.**

   You may sort more than two lists concurrently; just keep putting commas between the list names you enter.

7. **Press** ENTER **to sort the lists.**

   See the second screen in Figure 17-9.

8. **Press** STAT ENTER **to view list L₁.**

   See the third screen in Figure 17-9.

Chapter 18

Analyzing Statistical Data

. . . .

. . . .

*I*n descriptive statistical analysis, you usually want to plot your data and find the mean, median, standard deviation, and so on. You may also want to find a regression model for your data (a process also called *curve fitting*). This chapter explains how to get the calculator to do these things for you.

Plotting One-Variable Data

The most common plots used to graph one-variable data are histograms and box plots. In a *histogram,* the data is grouped into classes of equal size; a bar in the histogram represents one class. The height of the bar represents the quantity of data contained in that class, as in the first screen in Figure 18-1.

A *box plot* (as in the second screen in Figure 18-1) consists of a box-with-whiskers. The box represents the data existing between the first and third quartiles. The box is divided into two parts, with the division line defined by the median of the data. The endpoints of the whiskers represent the locations of the minimum and maximum data points. Sometimes, there are outliers at the end of the whiskers. Using your calculator, you can choose to show the outliers or include these points as part of the whiskers.

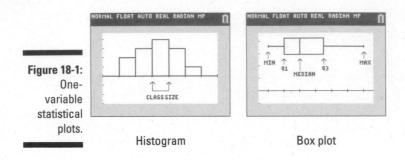

Figure 18-1:
One-
variable
statistical
plots.

Histogram   Box plot

Constructing a histogram

To construct a histogram of your data, follow these steps:

1. **Enter your data in the calculator.**

   See the first screen in Figure 18-2. Entering data in the calculator is described in Chapter 17. Your list does not have to appear in the Stat List editor to plot it, but it does have to be in the memory of the calculator.

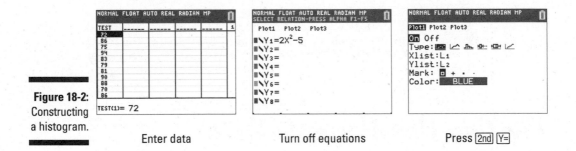

Figure 18-2:
Constructing
a histogram.

Enter data   Turn off equations   Press [2nd] [Y=]

2. **Turn off any Stat Plots or functions in the Y= editor that you don't want to be graphed along with your histogram.**

   To do so, press [Y=] to access the Y= editor. The calculator graphs any highlighted plots in the first line of this editor. To remove the highlight from a plot so that it won't be graphed, use the [▶] [◀] [▲] [▼] keys to place the cursor on the plot and then press [ENTER] to toggle the plot between highlighted and not highlighted.

   The calculator graphs only those functions in the Y= editor defined by a highlighted equal sign. To remove the highlight from an equal sign, use the [▶] [◀] [▲] [▼] keys to place the cursor on the equal sign in the definition of the function, and then press [ENTER] to toggle the equal sign between highlighted and not highlighted. See the second screen in Figure 18-2.

3. **Press ⌊2nd⌋⌊Y=⌋ to access the Stat Plots menu and enter the number (1, 2, or 3) of the plot you want to define.**

   The third screen in Figure 18-2 shows this process, where **Plot1** is used to plot the data.

4. **Highlight On.**

   If **On** is highlighted, the calculator is set to plot your data. If you want your data to be plotted at a later time, highlight **Off**. To highlight an option, use the ⌊▶⌋⌊◀⌋⌊▲⌋⌊▼⌋ keys to place the cursor on the option, and then press ⌊ENTER⌋.

5. **Press ⌊▼⌋, use ⌊▶⌋ to place the cursor on the type of plot you want to create, and then press ⌊ENTER⌋ to highlight it.**

   Select ⊡⊡ to construct a histogram.

6. **Press ⌊▼⌋, enter the name of your data list (Xlist), and press ⌊ENTER⌋.**

   If your data is stored in one of the default lists L_1 through L_6, press ⌊2nd⌋, key in the number of the list, and then press ⌊ENTER⌋. For example, press ⌊2nd⌋⌊1⌋ if your data is stored in L_1.

   If your data is stored in a user-named list, key in the name of the list and press ⌊ENTER⌋ when you're finished.

   You can always access a list by pressing ⌊2nd⌋⌊STAT⌋ and using the ⌊▲⌋⌊▼⌋ keys to scroll through all the stored lists in your calculator.

7. **Enter the frequency of your data.**

   If you entered your data without paying attention to duplicate data values, then the frequency is **1**. On the other hand, if you did pay attention to duplicate data values, you most likely stored the frequency in another data list. If so, enter the name of that list the same way you entered the **Xlist** in Step 6.

8. **Choose the color of your histogram.**

   Use the ⌊▶⌋⌊◀⌋ keys to operate the menu spinner to choose one of 15 color options. See the first screen in Figure 18-3.

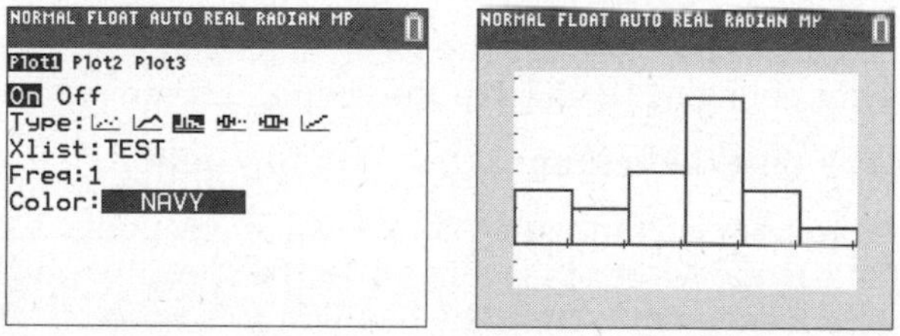

Figure 18-3:
Constructing
a histogram.

Stat Plot editor   Press ⌊ZOOM⌋ ⌊9⌋

9. **Press** ZOOM 9 **to plot your data using the ZoomStat command.**

ZoomStat finds an appropriate viewing window for plotting your data, as shown in the second screen in Figure 18-3. If you are not pleased with the graphing window that is generated, press WINDOW and change your values manually.

Adjusting the class size of a histogram

When creating a histogram, your calculator groups data into "classes." The data in the first screen in Figure 18-4 has been split into six classes represented by the six bars in the histogram.

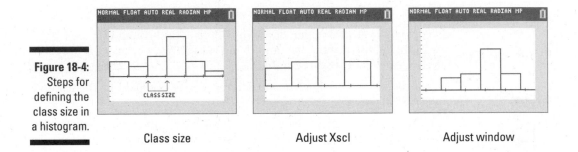

Figure 18-4:
Steps for defining the class size in a histogram.

Class size   Adjust Xscl   Adjust window

The class size (also called the class interval) is the width of each bar in the histogram. If you have more than 46 classes, your calculator will return the ERROR: STAT error message. Here is a formula that can be used to compute the class size:

Class size = (max – min)/(number of classes you want to have)

To adjust the class size of a histogram, follow these steps:

1. **Press** WINDOW, **set Xscl equal to the class size you desire, and then press** GRAPH.

To change the class size, change the value of **Xscl** in your calculator. See the graph after changing the **Xscl** in the second screen in Figure 18-4.

2. **If necessary, adjust the settings in the Window editor.**

When the histogram is graphed again using a different class size (as shown in the second screen in Figure 18-4), the viewing window doesn't do a good job of accommodating the histogram. To correct this, adjust the settings in the Window editor. I changed the Ymax settings to produce the third screen in Figure 18-4.

Constructing a box plot

To construct a box plot for your data, press [2nd][Y=][2] to access **Plot2**. Follow Steps 1 through 9 for constructing a histogram. In Step 5, select the Box Plot symbol ⊞ as shown in the first screen in Figure 18-5. If you adjust the viewing window, you can display a histogram and a box plot in the same viewing window (as shown in the second screen in Figure 18-5).

If your data has *outliers* (data values that are much larger or smaller than the other data values), consider constructing a modified box plot instead of a box plot. The third screen in Figure 18-5 illustrates both a standard box plot and a modified box plot of the same data. In a modified box plot, the whiskers represent data in the range defined by 1.5(Q3 − Q1), and the outliers are plotted as points beyond the whiskers. The steps for constructing box plots and modified box plots are the same, except in Step 5 you select the modified box plot symbol ⊞⋯.

Figure 18-5:
A box plot with a histogram and with a modified box plot.

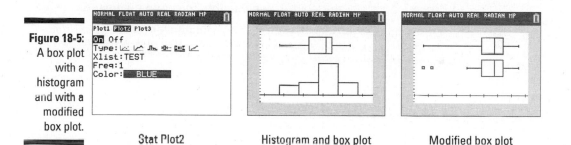

Stat Plot2   Histogram and box plot   Modified box plot

Plotting Two-Variable Data

The most common plots used to graph two-variable data sets are the scatter plot and the *xy*-line plot. The *scatter plot* plots the points (x, y), where x is a value from one data list (**Xlist**) and y is the corresponding value from the other data list (**Ylist**). The *xy-line plot* is simply a scatter plot with consecutive points joined by straight segments.

To construct a scatter plot or an *xy*-line plot, follow these steps:

1. **Follow Steps 1 through 6 in the previous section ("Constructing a histogram"), with the following difference:**

   In Step 5, highlight ⌐ to construct a scatter plot as shown in the first screen in Figure 18-6. Highlight to construct an *xy*-line plot.

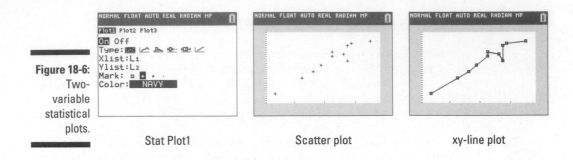

Figure 18-6:
Two-
variable
statistical
plots.

Stat Plot1   Scatter plot   xy-line plot

2. **Enter the name of your Ylist and press ENTER.**

3. **Choose the type of mark used to plot points.**

   You have four choices: a large empty square, a small plus sign, a small square, or a dot. To select one, use ▶ to place the cursor on the mark, and press ENTER.

4. **Press ZOOM 9 to plot your data using the ZoomStat command.**

   ZoomStat finds an appropriate viewing window for plotting your data. The second screen in Figure 18-6 shows a scatter plot, and the third screen in Figure 18-6 is an *xy*-line plot.

Tracing Statistical Data Plots

Before tracing a statistical data plot, press 2nd ZOOM and, if necessary, highlight **CoordOn** in the second line of the Format menu and **ExprOn** in the last line. This enables you to see the name of the data set being traced and the location of the cursor. To highlight an entry, use the ▶ ◀ ▲ ▼ keys to place the cursor on the entry and press ENTER.

Press TRACE to trace a statistical data plot. In the upper-left corner of the screen, you see the Stat Plot number (P1, P2, or P3) and the name(s) of the data list(s) being traced. If you have more than one stat plot on the screen, repeatedly press the ▲ ▼ keys until the plot you want to trace appears in the upper-left corner of the screen.

Use the ▶ ◀ keys to trace the plot. What you see depends on the type of plot:

  ✔ **Tracing a histogram:** As you trace a histogram, the cursor moves from the top center of one bar to the top center of the next bar. At the bottom of the screen, you see the values of **min**, **max**, and **n**. There are **n** data points *x* such that **min** ≤ *x* < **max**. This is illustrated in the first screen in Figure 18-7.

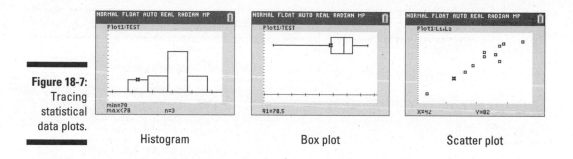

Figure 18-7:
Tracing
statistical
data plots.

Histogram   Box plot   Scatter plot

✔ **Tracing a box plot:** As you trace a box plot from left to right, the values that appear at the bottom of the screen are **minX** (the minimum data value), **Q1** (the value of the first quartile), **Med** (the value of the median), **Q3** (the value of the third quartile, and **maxX** (the maximum data value). This is illustrated in the second screen in Figure 18-7.

✔ **Tracing a modified box plot:** As you trace a modified box plot from left to right, the values that appear at the bottom of the screen are **minX** (the minimum data value), and then you see the values of the other out-liers, if any, to the left of the interval defined by 1.5(Q3 − Q1). The next value you see at the bottom of the screen is the value of the Left Bound of the interval defined by 1.5(Q3 − Q1). Then, as with a box plot, you see the values of the first quartile, the median, and the third quartile. After that you see the value of the right bound of the interval defined by 1.5(Q3 − Q1), the outliers to the right of this, if any, and finally, you see **maxX** (the maximum data value).

✔ **Tracing a scatter plot or an *xy*-line plot:** As you trace a scatter plot or an *xy*-line plot, the coordinates of the cursor location appear at the bottom of the screen. As shown in the third screen in Figure 18-7, the *x*-coordinate is a data value for the first data list named at the top of the screen; the *y*-coordinate is the corresponding data value from the second data list named at the top of the screen.

Analyzing Statistical Data

The calculator can perform one- and two-variable statistical data analysis. For one-variable data analysis, the statistical data variable is denoted by **x**. For two-variable data analysis, the data variable for the first data list is denoted by **x** and the data variable for the second data list is denoted by **y**. Table 18-1 lists the variables calculated using one-variable data analysis (**1-Var**), as well as those calculated using two-variable analysis (**2-Var**).

Table 18-1	One- and Two-Variable Data Analysis	
1-Var	*2-Var*	*Meaning*
$\bar{x}$	$\bar{x}, \bar{y}$	Mean of data values
Σx	$\Sigma x, \Sigma y$	Sum of data values
Σx^2	$\Sigma x^2, \Sigma y^2$	Sum of squares of data values
Sx	Sx, Sy	Sample standard deviation
σx	$\sigma x, \sigma y$	Population standard deviation
N	n	Total number of data points
Minx	minX, minY	Minimum data value
maxX	maxX, maxY	Maximum data value
Q1		First quartile
Med		Median
Q3		Third quartile
	Σxy	Sum of x*y

One-variable data analysis

To analyze one-variable data, follow these steps:

1. **Enter the data in your calculator.**

   Your list does not have to appear in the Stat List editor to analyze it, but it does have to be in the memory of the calculator.

2. **Press** [STAT] [▶] [1] **to activate the 1-Var Stats Wizard from the Stat Calculate menu.**

   See the first screen in Figure 18-8.

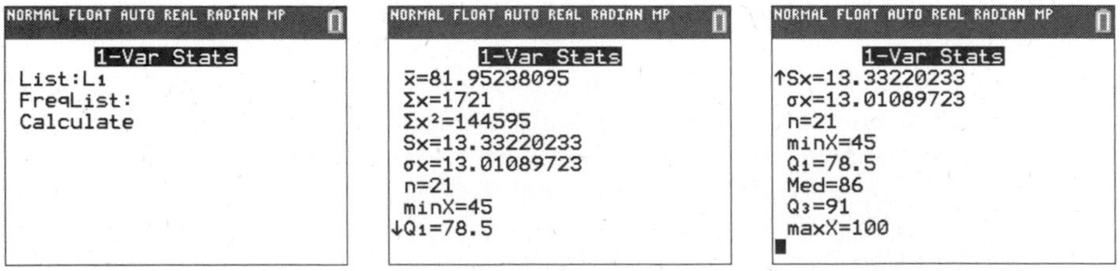

Figure 18-8:
Steps for one-variable data analysis.

NORMAL FLOAT AUTO REAL RADIAN MP	NORMAL FLOAT AUTO REAL RADIAN MP	NORMAL FLOAT AUTO REAL RADIAN MP
1-Var Stats	**1-Var Stats**	**1-Var Stats**
List:L₁	x̄=81.95238095	↑Sx=13.33220233
FreqList:	Σx=1721	σx=13.01089723
Calculate	Σx²=144595	n=21
	Sx=13.33220233	minX=45
	σx=13.01089723	Q₁=78.5
	n=21	Med=86
	minX=45	Q₃=91
	↓Q₁=78.5	maxX=100
Press [STAT] [▶] [1]	Press [ENTER]	Scroll results

3. **Enter the name of your data list (Xlist).**

   If your data is stored in one of the default lists L_1 through L_6, press 2nd, key in the number of the list, and then press ENTER. For example, press 2nd 1 if your data is stored in L_1.

   If your data is stored in a user-named list, press 2nd STAT, use the ▲▼ keys to scroll through all the stored lists in your calculator, and press ENTER to insert the list name you want.

4. **If necessary, enter the name of the frequency list.**

   If the frequency of your data is 1, you can skip this step and go to Step 5. If, however, you stored the frequency in another data list, enter the name of that frequency list (just as you entered the **Xlist** in Step 3).

5. **Press ENTER on CALCULATE to view the analysis of your data.**

   This is illustrated in the second screen in Figure 18-8. Use the ▲▼ keys to view the other values that don't appear on the screen. See the view after scrolling in the third screen in Figure 18-8.

Two-variable data analysis

To analyze two-variable data, follow these steps:

1. **Enter the data in your calculator.**

   Your data does not have to appear in the Stat List editor to analyze it, but it does have to be in the memory of the calculator.

2. **Press STAT ▶ 2 to activate the 2-Var Stats Wizard from the Stat Calculate menu.**

   See the first screen in Figure 18-9.

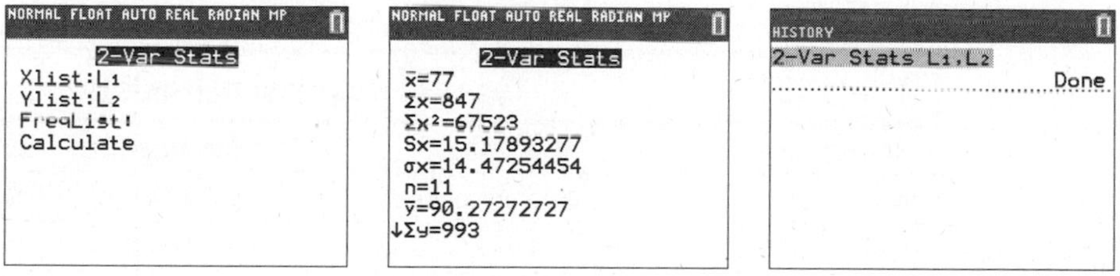

Figure 18-9:
Steps for
two-variable
data
analysis.

NORMAL FLOAT AUTO REAL RADIAN MP	NORMAL FLOAT AUTO REAL RADIAN MP	HISTORY
2-Var Stats	**2-Var Stats**	2-Var Stats L₁,L₂
Xlist:L₁	x̄=77	Done
Ylist:L₂	Σx=847	
FreqList!	Σx²=67523	
Calculate	Sx=15.17893277	
	σx=14.47254454	
	n=11	
	ȳ=90.27272727	
	↓Σy=993	

Press STAT ▶ 2   Press ENTER   Scroll history

3. **Enter the name of your data list (Xlist).**

If your data is stored in one of the default lists L_1 through L_6, press 2nd, key in the number of the list, and then press ENTER. For example, press 2nd1 if your data is stored in L_1.

If your data is stored in a user-named list, press 2ndSTAT, use the ▲▼ keys to scroll through all the stored lists in your calculator, and press ENTER to insert the list name you want.

4. **Enter the name of the Ylist.**

5. **If necessary, enter the name of the frequency list.**

If the frequency of your data is 1, you can skip this step and go to Step 6. If, however, you stored the frequency in another data list, enter the name of that frequency list (just as you entered the **Xlist** in Step 3).

6. **Press ENTER on CALCULATE to view the analysis of your data.**

This is illustrated in the second screen in Figure 18-9. Use the ▲▼ keys to view the other values that don't appear on the screen.

If you press 2ndMODE to dismiss your results, you will find your cursor on a clear Home screen. Press ▲ to scroll through your previous calculations and you will find the 2-Var Stats command as shown in the third screen in Figure 18-9. If you plan on using the 2-Var Stats command multiple times, save time by highlighting and pressing ENTER to paste the command into your current entry line.

Performing regressions

Regression modeling is the process of finding a function that approximates the relationship between the two variables in two data lists. Table 18-2 shows the types of regression models the calculator can compute.

Table 18-2	Types of Regression Models	
TI-Command	*Model Type*	*Equation*
Med-Med	Median-median	$y = ax + b$
LinReg(ax+b)	Linear	$y = ax + b$
QuadReg	Quadratic	$y = ax^2 + bx + c$
CubicReg	Cubic	$y = ax^3 + bx^2 + cx + d$
QuartReg	Quartic	$y = ax^4 + bx^3 + cx^2 + dx + e$
LinReg(a+bx)	Linear	$y = a + bx$
LnReg	Logarithmic	$y = a + b*\ln(x)$

TI-Command	Model Type	Equation
ExpReg	Exponential	$y = ab^x$
PwrReg	Power	$y = ax^b$
Logistic	Logistic	$y = c/(1 + a*e^{-bx})$
SinReg	Sinusoidal	$y = a*\sin(bx + c) + d$

To compute a regression model for your two-variable data, follow these steps:

1. **If necessary, turn on Diagnostics and put your calculator in Function mode.**

   When Stat Diagnostics is turned on, the calculator displays the correlation coefficient (r) and the coefficient of determination (r^2 or R^2) for appropriate regression models (as shown in the third screen in Figure 18-10). By default, Stat Diagnostics is turned off.

   If the regression model is a function that you want to graph, you must first put your calculator in Function mode.

   Here's how to turn Stat Diagnostics on and set your calculator to Function mode:

   a. Press MODE.

   b. Use the ▶◀▲▼ *keys to highlight STAT DIAGNOSTICS ON and press* ENTER.

   c. Use the ▶◀▲▼ *keys to highlight FUNCTION and press* ENTER.

   The first screen in Figure 18-10 shows this procedure.

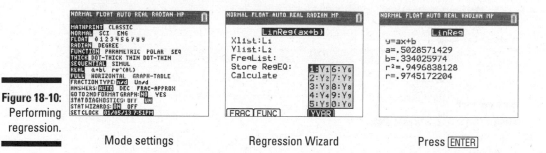

Figure 18-10: Performing regression.

Mode settings   Regression Wizard   Press ENTER

2. **Select a regression model from the Stat CALCULATE menu to activate the Regression Wizard.**

   Press STAT ▶ to access the Stat CALCULATE menu. Repeatedly press ▼ until the number or letter of the desired regression model is highlighted, and press ENTER to select that model.

3. **Enter the name for the Xlist data and enter the name of the Ylist data.**

   The appropriate format for entering list names is explained in Step 3 in the earlier section, "One-variable data analysis." The default Xlist and Ylist are L_1 and L_2.

4. **If necessary, enter the name of the frequency list.**

5. **With your cursor in the Store RegEQ line, enter the name of the function (Y_1, ... , Y_9, or Y_0) in which the regression model is to be stored.**

   To enter a function name, press [ALPHA][TRACE] to access the shortcut Y-VAR menu and then enter the number of the function you want, as shown in the second screen in Figure 18-10.

6. **Press [ENTER] on CALCULATE to view the equation of the regression model.**

   This is illustrated in the third screen in Figure 18-10. The equation of the regression model is automatically stored in the Y= editor under the name you entered in Step 5.

Graphing a regression model

Often, it is a good idea to take a look at the scatter plot of your data to determine what type of regression model is best. Here are the steps to graph a scatter plot of your data and the regression model on the same graph:

1. **If you haven't already done so, graph your two-variable data in a scatter plot or an *xy*-line plot.**

   Set up the scatter plot by pressing [2nd][Y=][ENTER]. See Stat Plot1 in the first screen in Figure 18-11. The earlier section in this chapter, "Plotting Two-Variable Data," explains how to do so.

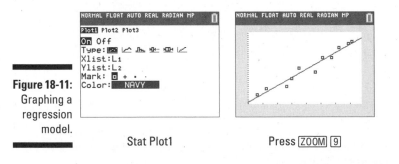

Figure 18-11:
Graphing a
regression
model.

Stat Plot1　　　　　Press [ZOOM] [9]

2. **Press [ZOOM][9] to see the graph of your data and regression model.**

   This process is illustrated in the second screen in Figure 18-11.

Graphing a residual plot

A residual plot shows the residuals on the vertical axis and the independent variable on the horizontal axis. What are residuals? Residuals are a sum of deviations from the regression line. Because a linear regression is not always the best choice, residuals help you figure out if your regression model is a good fit for your data. Here are the steps to graph a residual plot:

1. **Press** Y= **and deselect stat plots and functions.**

   To remove the highlight from a plot so that it won't be graphed, use the ▶◀▲▼ keys to place the cursor on the Plot and then press ENTER.

   To remove the highlight from an equal sign, use the ▶◀▲▼ keys to place the cursor on the equal sign in the definition of the function, and then press ENTER.

2. **Press** 2nd Y= 2 **to access Stat Plot2 and enter the Xlist you used in your regression.**

3. **Enter the Ylist by pressing** 2nd STAT **and using the** ▲▼ **keys to scroll to RESID.**

   See the first screen in Figure 18-12.

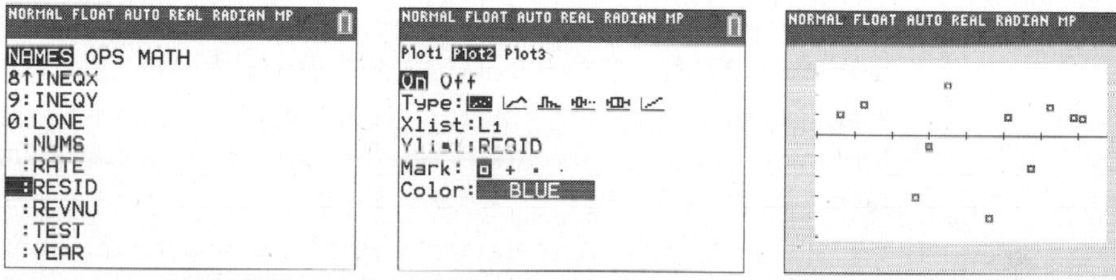

Figure 18-12: Graphing a residual plot.

RESID list   Stat Plot2   Press ZOOM 9

4. **Press** ENTER **to insert the RESID list.**

   See the second screen in Figure 18-12.

5. **Press** ZOOM 9 **to graph the residual plot.**

   See the third screen in Figure 18-12.

Using Manual-Fit

Do you think you could come up with a better line of best fit than your calculator did with its regression line? Go ahead and try! Manual Linear Fit enables you to visually find a line of best fit of the form **Y=mX+b**. Here are the steps for using Manual Linear Fit:

1. **Press** Y= **and deselect any functions that would graph by pressing** ENTER **on the corresponding equal sign.**

2. **Press** STAT ▶ ▲ ▲ ENTER **to open the Manual-Fit Wizard.**

   Manual-Fit is located near the bottom of the Stat CALC menu. On the TI-84 Plus, it is the last entry in this menu.

3. **With your cursor on Store EQ, press** ALPHA TRACE **to access the shortcut Y-VAR menu.**

   See the first screen in Figure 18-13. Enter the number of the Y-VAR you want. Press ▼ to Highlight CALCULATE and press ENTER.

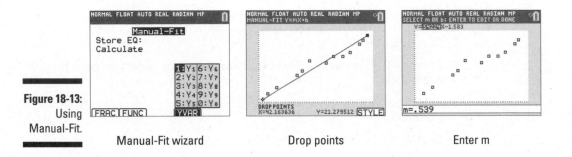

Figure 18-13: Using Manual-Fit.

Manual-Fit wizard   Drop points   Enter m

4. **Use the** ▶ ◀ ▲ ▼ **keys to navigate your cursor near the data point closest to the right part of the screen and press** ENTER.

   Two points determine a line; this establishes a point on the Manual-Fit line. As you move your cursor, the Manual-Fit line behaves like a moveable line as shown in the second screen in Figure 18-13.

5. **Use the** ▶ ◀ ▲ ▼ **keys to navigate your cursor near the data point closest to the left part of the screen and press** ENTER.

   The Manual-Fit line is now drawn with the equation shown in the border at the top of the graph screen.

   The TI-84 Plus C displays functions and information in the border of the graph screen. The TI-84 Plus displays similar information directly on the graph screen.

6. **Use the** ▶ ◀ **keys to toggle the highlighted parameter values from** *m* **to** *b* **in the Manual-Fit line equation of the form, Y=*m*X+*b*.**

7. **Enter a value for the highlighted parameter value to adjust the fit of your line.**

   Notice, an entry line opens at the bottom of the screen as you enter a value for one of the parameters. See the third screen in Figure 18-13. Press ENTER to change the parameter in the equation and watch the graph automatically adjust.

8. **Press** GRAPH **to activate the on-screen prompt, DONE and then press** 2nd MODE **to exit the graph screen.**

This action stores the function and brings your cursor to the Home screen. On the TI-84 Plus, press 2nd MODE to store the function.

Using statistics commands on the Home screen

I want to show you a few more statistic commands. Press 2nd STAT ◄ to access the Stat List MATH menu, as shown in the first screen in Figure 18-14. For example, you can quickly calculate the mean of a short list of numbers. On a Home screen, press 2nd STAT ◄ 3 to insert the **mean(** command. Press 2nd (and enter a list of numbers separated by commas, or press 2nd 1 to insert list L_1. See the second screen in Figure 18-14.

You can use the 1-Var Stats Wizard (STAT ► 1) to calculate the min, max, Q1, Q3, median, mean, and standard deviation of a data list.

If you are taking a statistics course, check out the Stat TESTS menu by pressing STAT ◄ as shown in the third screen in Figure 18-14.

Figure 18-14: Using statistics commands on the Home screen.

NORMAL FLOAT AUTO REAL RADIAN MP
NAMES OPS **MATH**
1:min(
2:max(
3:mean(
4:median(
5:sum(
6:prod(
7:stdDev(
8:variance(

Press 2nd STAT ◄

NORMAL FLOAT AUTO REAL RADIAN MP
mean({2,3,3,6,9})
4.6
mean(L₁)
74

mean(command

NORMAL FLOAT AUTO REAL RADIAN MP
EDIT CALC **TESTS**
1:Z-Test…
2:T-Test…
3:2-SampZTest…
4:2-SampTTest…
5:1-PropZTest…
6:2-PropZTest…
7:ZInterval…
8:TInterval…
9↓2-SampZInt…

Press STAT ◄

Part V
Doing More with Your Calculator

In this part . . .

- ✔ Use the built-in Finance app to calculate how much money you need to save to become a millionaire!

- ✔ Learn how to download and install TI Connect software so you can (among other things) transfer files to and from your PC.

- ✔ Learn to transfer files from your calculator to another (or vice versa).

- ✔ See how to insert images on a graph, plot points, and perform a regression right on a graph.

- ✔ Discover the archive, group, and reset features to preserve memory on your calculator

Chapter 19

Using the TVM Solver

*D*o you understand how your money works? Money can work against you when you take on a loan or mortgage. Money can also work for you when you save and invest money. Using the built-in functionality of the TVM Solver, you can learn to set savings goals and use the power of compound interest to help you reach your goals. I love using the TVM Solver to figure out how much I need to save in order to become a millionaire! If that sounds interesting, keep reading.

Calculating Mortgages and Loans

Before you start using the TVM Solver, you need to know a few of the basics. Here is a list of TVM variables:

✔ **N:** Total number of payments. An easy way to calculate this is to multiply the **P/Y** times the number of years.

✔ **I%:** Annual interest rate. Always enter this rate as a percentage!

✔ **PV:** Present value. This is how much you invest or the amount of the loan.

✔ **PMT:** Payment amount.

✔ **FV:** Future value. This is your savings goal or if you are calculating loans, your loan balance at the end of the loan.

✔ **P/Y:** Number of payment periods per year.

✔ **C/Y:** Number of compounding periods/year.

In this example, I use the TVM Solver to answer this question:

The average home price in Memphis, Tennessee, is $154,985. If the interest rate on a 30-year loan in 4.73%, what will the monthly payment be?

Step 1. Access the TVM Solver

Follow these steps to access the TVM Solver:

1. **Press APPS to access the apps that are loaded on your calculator.**

See the first screen in Figure 19-1.

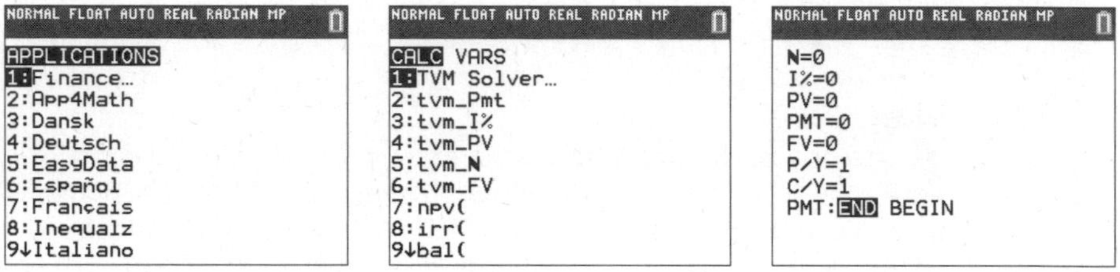

Figure 19-1: Accessing the TVM Solver.

Press APPS   Finance app   TVM Solver

2. **Press 1 or ENTER to start the Finance app.**

See the second screen in Figure 19-1.

3. **Press 1 or ENTER to display the TVM Solver.**

See the third screen in Figure 19-1.

Step 2. Enter values for five of the six TVM variables

When entering values, you must keep in mind the cash flow convention. If you are getting money (like a mortgage or loan), values are entered as positive numbers. If you are investing money, values are entered as negative numbers. The cash flow convention could affect the sign of these three TVM variables: PV, PMT, and FV.

I consider P/Y and C/Y to be the same variable. If you enter a value for P/Y, C/Y automatically changes to match. You can change the value of C/Y to make it a unique value, but in most cases, these two variables end up being identical. To use the TVM Solver, enter values for five of the six TVM variables with the information that was given to you in the problem presented.

Here are steps to enter values for five of the variables.

1. **Enter N.**

   How many mortgage payments are there over the life of the loan? Twelve payments a year for 30 years. If you enter the expression **12*30**, your calculator will evaluate the expression when you use the ⏷ key or press ⏎ENTER⏎ to move to the next field.

2. **Enter I%.**

   4.73% is entered as 4.73.

3. **Enter PV.**

   The present value in this problem is the amount of money of the loan; $154,985 is entered as 154985.00.

4. **Enter PMT.**

   Isn't this what you're trying to find? You aren't allowed to leave a field blank, so leave the default value of 0.

5. **Enter FV.**

   The future value is the amount of money you still have to pay after the 30-year term, which is 0.

6. **Enter P/Y and C/Y.**

   Mortgages have monthly payments and are compounded monthly, so enter 12.

7. **Highlight PMT: END or BEGIN.**

   This problem does not specify if the payment is at the beginning or end of the month. I usually assume payments are at the end of the month (the default setting). Use the ▶◀ keys to move your cursor to END and press ⏎ENTER⏎ to highlight. See the first screen in Figure 19-2.

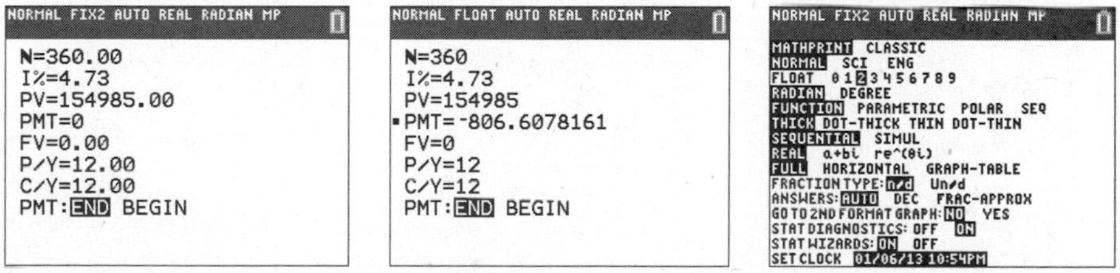

Figure 19-2:
Using the
TVM Solver.

| Enter values | Press ⏎ALPHA⏎ ⏎ENTER⏎ | Press ⏎MODE⏎ |

Step 3. Solve for the missing TVM variable

Look back at what the question is asking: "What will the monthly payment be?"

1. **Use the ⌃⌄ keys to place your cursor on the value you would like to solve for and press** ALPHA ENTER.

   Move the cursor to the PMT value and press ALPHA ENTER. See the second screen in Figure 19-2. Notice the small box next to the PMT value, indicating that you solved for that variable. Why is the value negative? Paying a mortgage is a cash outflow, so it must be negative according to the cash flow convention.

   Before using TVM Solver, it may be a good idea to press MODE to change the mode of your calculator to Float to 2 as shown in the third screen in Figure 19-2. This will automatically round any values to the hundredths place. Since the TVM solver works with monetary values, everything will be rounded to the nearest penny.

Graphing an amortization table

An amortization schedule shows the amount of interest and principal for each periodic payment made over the life of a loan. An amortization schedule doesn't usually fit on one page. Wouldn't it be great to see a visual snapshot of a loan amortization showing the remaining balance for each period of the loan?

The easiest way to graph an amortization table is by using a parametric graph. To graph an amortization of the problem posed in the last section, press MODE and change the mode to PARAMETRIC. Press Y= and enter X_{1T}=**T**. In

Parametric mode, press X,T,Θ,n to enter T. Enter Y_{1T}=**bal(T)**. To insert the balance command, press APPS ENTER 9 to insert the **bal(** command. See the first screen in the figure. Press WINDOW and configure the window variables as shown in the second screen in the figure. Press GRAPH to view the graph of the amortization. The *x*-value of the graph is the payment number and the *y*-value of the graph is the remaining balance. Press TRACE to take a closer look at your graph as shown in the third screen in the figure.

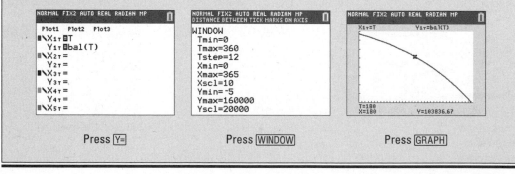

Press Y=   Press WINDOW   Press GRAPH

Making Compound Interest Work for You

Hopefully, you are beginning to get the hang of entering values into the TVM Solver. It might be a good idea to get a little more practice. This time, you get to see what happens when compound interest is working for you rather than against you.

In this example, use the TVM Solver to answer this question:

> You make a new year's resolution to give up your daily trip to the coffee shop where you spend about $5 a day. Instead, you decide to save the money in a cookie jar and make an end-of-the-year investment of $1,825 in a growth stock mutual fund. If the mutual fund gets a 10% return a year (consider it compounded yearly), how long will it take for you to become a millionaire?

Step 1. Access the TVM Solver

Follow this step to access the TVM Solver:

1. **Press** APPS ENTER ENTER **to access the TVM Solver.**

Step 2. Enter values for five of the six TVM variables

Here are the steps to enter values for five of the variables.

1. **Enter N.**

   How long will it take to become a millionaire? That is the question you are trying to answer. You aren't allowed to leave a field blank, so just leave the default value of 0.

2. **Enter I%.**

   10% is entered as 10.

3. **Enter PV.**

   The present value in this problem is 0.

4. **Enter PMT.**

Remember, a cash outflow is considered a negative value. Enter the amount you are investing each year as a negative number, –1825. Alternatively, you could enter the expression **–(5*365)** and your calculator will evaluate the expression when you use the ⊡ key or press ⌷ENTER⌷ to advance to the next field.

5. **Enter FV.**

   The future value is your financial goal, $1,000,000. Notice, this number is entered as a positive number because it is a cash inflow. Of course, don't use commas or the $ sign when entering the number.

6. **Enter P/Y and C/Y.**

   Enter **1** for the number of payments per year.

7. **Highlight PMT: END or BEGIN.**

   Payment is made at the end of the year in this problem. Use the ▶◀ keys to move your cursor to END, and press ⌷ENTER⌷ to highlight. See the first screen in Figure 19-3.

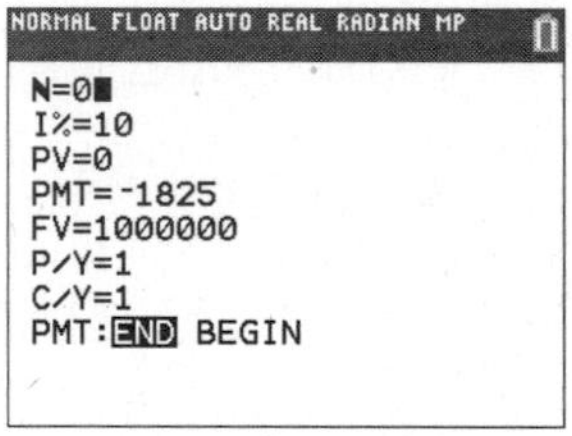

Figure 19-3:
Using the
TVM Solver.

NORMAL FLOAT AUTO REAL RADIAN MP
N=0▪
I%=10
PV=0
PMT=-1825
FV=1000000
P/Y=1
C/Y=1
PMT:**END** BEGIN

Enter values

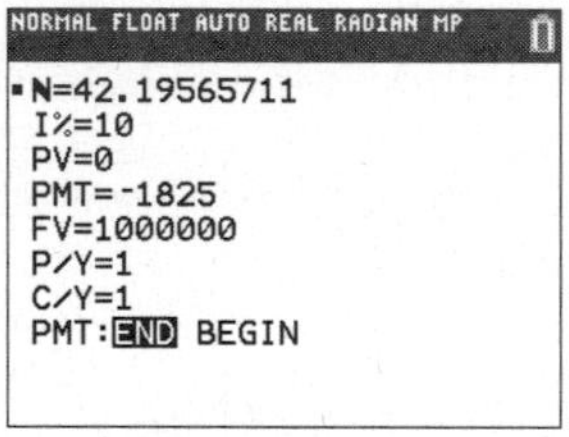

NORMAL FLOAT AUTO REAL RADIAN MP
▪N=42.19565711
I%=10
PV=0
PMT=-1825
FV=1000000
P/Y=1
C/Y=1
PMT:**END** BEGIN

Press ⌷ALPHA⌷ ⌷ENTER⌷

Step 3. Solve for the missing TVM variable

Look back at what the question is asking: "How long will it take to become a millionaire?"

1. **Use the** ▲▼ **keys to place your cursor on the value you would like to solve for and press** ⌷ALPHA⌷⌷ENTER⌷.

   Move the cursor to the N value and press ⌷ALPHA⌷⌷ENTER⌷. See the second screen in Figure 19-3. Notice the small box next to the N value, indicating that you solved for that variable.

Were you surprised at the result? In a little over 42 years, you could be a millionaire!

Chapter 20

Communicating with a PC Using TI Connect Software

*I*n this chapter, I explain how to use TI Connect Software to transfer files between your calculator and your PC. Of course, first I show you how to download the software. As an added benefit to installing TI Connect Software, a USB driver is installed that enables you to recharge the TI-84 Plus C using your computer! Keep reading to find all that you can do when a computer can communicate with your calculator.

You need two things to enable your calculator to communicate with your computer: TI Connect Software and either a USB computer cable or a USB Silver Edition Cable. TI Connect is free, and the cable came bundled with your calculator. If you are no longer in possession of the cable, you can purchase one through the Texas Instruments online store at http://education.ti.com.

Downloading TI Connect

The following steps explain how to download the current version of TI Connect from the Texas Instruments website, as it existed at the time this book was published:

1. **Go to the Texas Instruments website (**http://education.ti.com**).**

2. **Locate the Downloads drop-down menu and select Apps, Software & Updates.**

3. **Under the Technology drop-down menu, select TI-84 Plus Family, TI-83 Plus Family.**

4. **Click the Find button, scroll down, and select TI Connect Software.**

5. **Select the appropriate language.**

6. **Follow the directions given during the downloading process. Make a note of the directory in which you save the download file.**

You can download an extensive TI Connect Help document (more than 100 pages) by clicking the Help icon in the bottom-right corner of the TI Connect Home screen.

Installing and Running TI Connect

After you've downloaded TI Connect, you install it by double-clicking the downloaded TI Connect file you saved on your computer. Then follow the directions given by the TI Connect Installation Wizard.

When you start the TI Connect program, you see the many subprograms it contains. See the TI Connect Home screen in Figure 20-1.

A USB driver is automatically installed on your computer when you download and install TI Connect software. Now, you can recharge the battery on the TI-84 Plus C when you use the USB computer cable to connect your calculator with a computer.

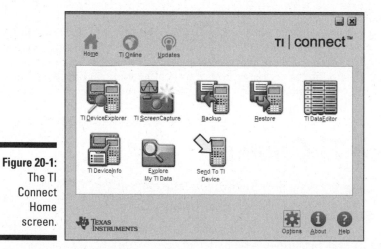

Figure 20-1:
The TI Connect Home screen.

Here is a short description of the TI Connect software tools:

- ✔ **TI Device Explorer:** Transfer files between your calculator and a computer.

- ✔ **TI Screen Capture:** Capture screenshots from your calculator. Chapter 22 explains how to use this tool to transfer an image from your computer to your calculator.

- ✔ **Backup:** Back up the files stored on your calculator.

- ✔ **Restore:** Restore calculator files that have been previously backed up.

- ✔ **TI Data Editor:** Create or edit data variables.

- ✔ **TI Device Info:** Find information about connected calculators.

- ✔ **Explore My TI Data:** Opens Windows Explorer to work with calculator files.

- ✔ **Send to TI Device:** Opens a dialog box you can use to transfer files to your calculator.

Would you like to have the most current version of TI Connect? If so, click the Options icon on the TI Connect Home screen and select the box that says, "Check for new software on every startup."

Connecting Calculator and PC

You can use the cable that came with your calculator to connect your calculator to your computer.

The USB computer cable that came with your calculator is a USB-to-mini-USB cable. Because the ends of this cable are of different sizes, it's easy to figure out how to connect your calculator to your computer; the small end fits in the right slot on the top of your calculator and the other end plugs into one of your computer's USB ports.

The USB Silver Edition Cable can be used to connect your calculator to your computer. The plug end of this cable fits into the top left slot on your calculator, called the I/O port.

Press ON after you connect your calculator to your computer using a USB computer cable. The action of turning on your calculator helps your computer recognize the device that is connected through the USB hub of your computer.

Click the TI Device Explorer icon on the TI Connect Home screen. A Select TI Device dialog box opens. It may take a minute of searching to find your connected device. When you see your device in the dialog box, click on it to highlight and then click the Select button as shown in Figure 20-2.

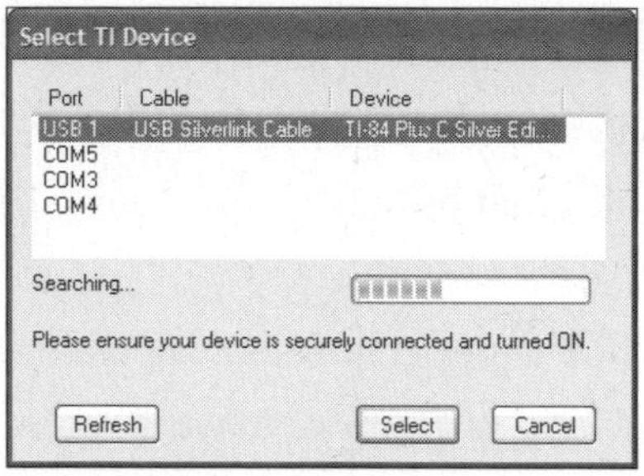

Figure 20-2:
The Select
TI Device
dialog box.

Transferring Files

TI Connect software can be used to transfer files between a computer and your calculator. The directions for sending files *from* or *to* a calculator are a bit different, so I provide separate sections of instructions for each type of transfer.

Using Device Explorer to transfer files from calculator to computer

After you've connected the calculator to your computer, the TI Device Explorer program housed in TI Connect can transfer files between the two devices. To transfer files between your calculator and PC, start the TI Connect software and click the TI Device Explorer program. A directory appears, listing the files on your calculator. Expanding this directory works the same on your calculator as on your computer. When transferring files, you're usually interested in transferring the files housed in the following directories: Graph Database, List, Matrix, Background (Picture and Image), and Program. If any of these directories don't appear on-screen, that means no files are housed in that directory.

1. **Highlight the files you want to transfer.**

   Hold the shift key on your computer to highlight consecutive files, and hold the control key on your computer to highlight multiple files not listed consecutively. See Figure 20-3.

Figure 20-3:
Transferring
files from
your cal-
culator to a
computer.

2. **Click File and select either Copy to PC or Move to PC.**

   Copy to PC will place a copy of the file on the computer. Move to PC will delete the file from your calculator and move the file to the computer.

3. **Select the location for your files in the Choose Folder window and click OK.**

The Device Explorer window can be used to drag and drop files from your computer to your calculator, or vice versa. Just open a computer documents folder and a Device Explorer window at the same time and let the dragging and dropping fun begin!

Using Send to TI Device to transfer files from computer to calculator

To copy files to the calculator from a PC running Windows, you don't need to be in the TI Device Explorer window. Just open Windows Explorer, highlight the files you want to copy, right-click the highlighted files, and select Send To TI Device. Your files are populated in the Send To TI Device window. You have all kinds of options, including setting the file to RAM or Archive. See Figure 20-4. When you are ready to send, click Send to Device.

Directions for transferring files from a Macintosh to the calculator can be found in the TI Device Explorer Help menu.

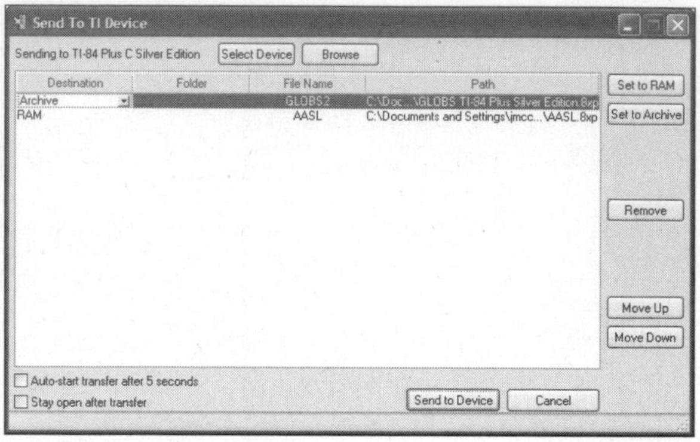

Figure 20-4:
Transferring
files from a
computer
to your
calculator.

Using the Backup and Restore Tools

If your rechargeable battery loses its charge, you are in real danger of losing the data stored in RAM. If your friend borrows your calculator and accidentally deletes all memory (I have seen this happen), how can you get your data back? I am guessing that you already understand the importance of backing up the files on your computer. If you have calculator files that would be hard to replace, you need to learn how to use the Backup and Restore tools.

The Backup tool can be accessed by clicking the Backup icon on the TI Connect Home screen. The Backup dialog box opens on the computer screen, as shown in Figure 20-5. Select the data you want to back up and choose the location on your computer where you want the data to be stored. The default location for your Backup data is: My Computer⇨My TIData⇨Backups. Click OK and wait for the selected data to be stored on the computer.

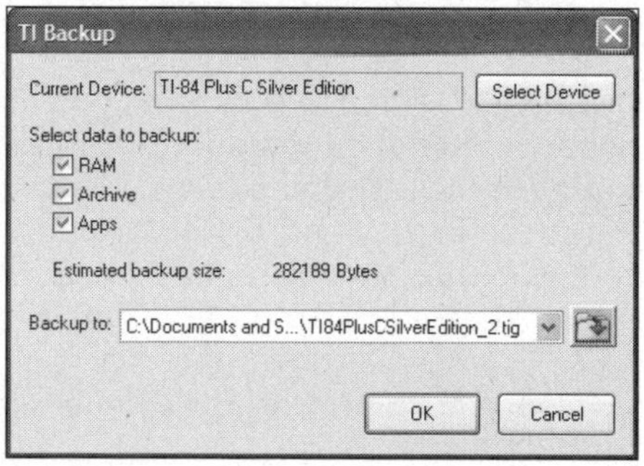

Figure 20-5:
Backing
up data.

The Restore tool can be accessed by clicking the Restore icon on the TI Connect Home screen. When the Restore icon is clicked, a window opens, prompting you to select the location where the Backup data was saved on your computer. After you select the location, you can select the data you want to restore in the Restore dialog box. Click OK to restore the data on your calculator.

Upgrading the OS

Texas Instruments periodically upgrades the operating systems of the TI-84 Plus C and TI-84 Plus families of calculators. To get the calculator to upgrade the operating system, start the TI Connect software and click the Updates icon on the TI Connect Home screen. Select Device Operating System in the TI Updates dialog box as shown in Figure 20-6. Click Continue and follow on-screen directions to update the OS on your calculator.

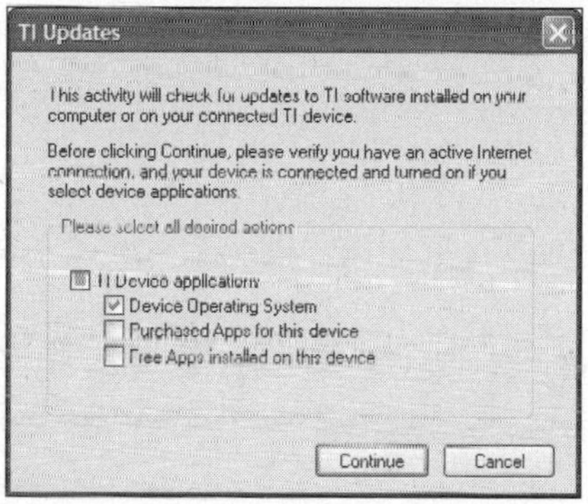

Figure 20-6:
The TI Updates dialog box.

Chapter 21

Communicating Between Calculators

. . . .

In This Chapter

▶ Linking calculators so files can be transferred between them

▶ Determining what files can be transferred

▶ Selecting files to be transferred

▶ Transferring files between calculators

. . . .

*Y*ou can transfer data lists, programs, matrices, and other such files from one calculator to another if you link the calculators with the unit-to-unit link cable that came with your calculator. This chapter describes how to make such transfers.

Linking Calculators

Calculators are linked using the unit-to-unit link cable that came with the calculator. If you are no longer in possession of the cable, you can purchase one through the Texas Instruments online store at http://education.ti.com.

The unit-to-unit link cable has an I/O plug on each end. This cable can be used to link a TI-83 to a TI-84. It can also be used to link two TI-84s. On the TI-83, it connects to the I/O port at the bottom of the calculator.

The unit-to-unit USB cable that came with your calculator has two mini-USB connectors. This cable can be used to connect TI-84 calculators using the mini-USB port at the top of each calculator.

If you get an error message when transferring files from one calculator to another, the most likely cause is that the unit-to-unit USB cable isn't fully inserted into the port of one calculator.

Transferring Files

You can transfer files between any of the TI-84 Plus C, TI-84 Plus, and TI-83 families of calculators. After connecting two calculators, you can transfer files from one calculator (the sending calculator) to another (the receiving calculator).

TI-84 Plus C calculator files are usually, but not always, compatible with the TI-84 Plus and TI-83 families of calculators. Apps, pics, and images are not compatible files. Most calculator programs will transfer, but functionality differences may cause problems in the execution of the program. If you try to transfer an incompatible calculator file, you will get the ERROR: VERSION error message as shown in the first screen in Figure 21-1.

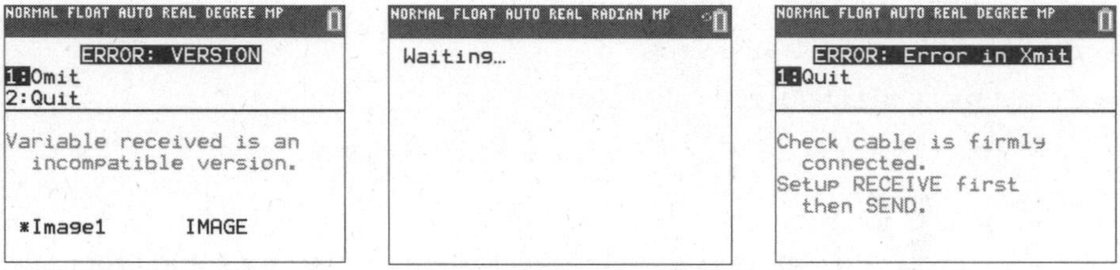

Figure 21-1:
Selecting
files for
trans-
mission
between
calculators.

NORMAL FLOAT AUTO REAL DEGREE MP

    ERROR: VERSION
1:Omit
2:Quit

Variable received is an
 incompatible version.

∗Image1   IMAGE

ERROR: VERSION

NORMAL FLOAT AUTO REAL RADIAN MP

Waiting...

Receiving calculator

NORMAL FLOAT AUTO REAL DEGREE MP

    ERROR: Error in Xmit
1:Quit

Check cable is firmly
 connected.
Setup RECEIVE first
 then SEND.

Error in Xmit

To select and send files, follow these steps:

1. **On the receiving calculator, press** 2nd X,T,Θ,n ▶ ENTER.

   You see a screen that says **Waiting**, and in its upper-right corner, a moving dashed line indicates that the receiving calculator is waiting to receive files. See the second screen in Figure 21-1.

   Always put the receiving calculator in Receiving mode *before* you transfer files from the sending calculator! If you forget, you will get the ERROR: Error in Xmit error message as shown in the third screen in Figure 21-1. Notice the helpful messages at the bottom of the error message screen.

2. **Press** 2nd X,T,Θ,n **on the sending calculator to access the Link SEND menu.**

   See the first screen in Figure 21-2.

3. **Use the** ▲▼ **keys to select the type of file you want to send, and then press** ENTER.

The first screen in Figure 21-2 shows the types of files you can send. The down arrow visible after number 9 in this list of menu items indicates that there are more menu items than can be displayed on-screen. Press ⌐ to view these other menu items.

If you want to send all files on the calculator to another calculator, select **All+** and proceed to Step 4. If you select **All–**, you have the opportunity to select the files you want to send from an exhaustive list of every file that could possibly be sent!

4. **Use the ⌐⌐ keys to move the cursor to a file you want to send, and press ENTER to select that file. Repeat this process until you have selected all the files in this list that you want to send to another calculator.**

The calculator places a small square next to the files you select, as in the second screen in Figure 21-2. In this screen, lists L_1, L_3, and L_4 are selected in the **List SELECT** menu.

Figure 21-2:
Selecting and receiving calculator screens.

NORMAL FLOAT AUTO REAL RADIAN MP	NORMAL FLOAT AUTO REAL RADIAN MP	NORMAL FLOAT AUTO REAL RADIAN MP
SEND RECEIVE	SELECT TRANSMIT	SELECT TRANSMIT
1:All+…	▪ L₁ LIST	1:Transmit
2:All–…	L₂ LIST	
3:Prgm…	▪ L₃ LIST	
4:List…	▪ L₄ LIST	
5:GDB…	▶ L₅ LIST	
6:Pic & Image…	L₆ LIST	
7:Matrix…	DEP LIST	
8:Real…	INEQX LIST	
9↓Complex…	INEQY LIST	

Press 2nd X,T,Θ,n   Select items   Sending calculator

5. **On the sending calculator, press ▶ to access the Link TRANSMIT menu.**

See the third screen in Figure 21-2.

6. **On the sending calculator, press ENTER to send the files to the receiving calculator.**

As files are transferred, you may receive the **DuplicateName** menu, as illustrated in the first screen in Figure 21-3. This indicates that the receiving calculator already contains a file with the same name. Because the default names for stat lists are stored in the calculator, you always get this message when transferring L_1–L_6, even if the list on the receiving calculator is empty.

When you get the **DuplicateName** menu, select the appropriate course of action:

• If you select **Overwrite**, any data in the existing file is overwritten by the data in the file being transferred.

• If you select **Rename**, a new file is created and stored under the name you specify, as in the second screen in Figure 21-3.

The Ⓐ after **Name =** indicates that the calculator is in Alpha mode: When you press a key, the green letter above the key is displayed. To enter a number, press ALPHA to take the calculator out of Alpha mode and then enter the number. To enter a letter after entering a number, you must first press ALPHA. Press ENTER when you are finished entering the name.

When renaming a file that is being transferred to the receiving calculator, the calculator has a strange and confusing way of warning you if you already have a file on the receiving calculator with that name. When you press ENTER after entering the name, the calculator erases the name and makes you start over entering a name. No warning message tells you that a file having the same name already exists on the calculator. If this happens to you, simply enter a different name.

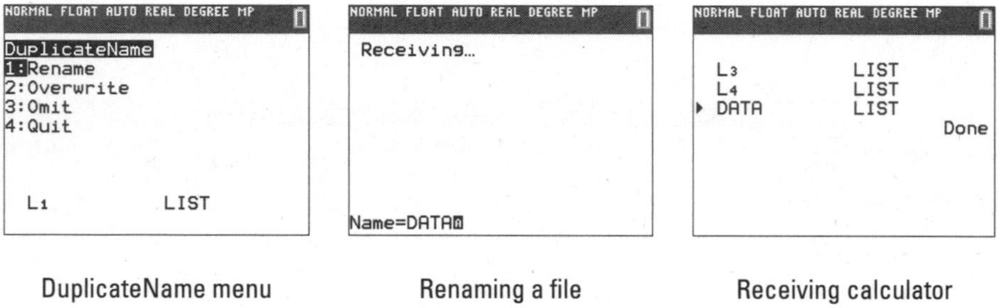

Figure 21-3:
Dealing with
duplicate
file names.

| DuplicateName menu | Renaming a file | Receiving calculator |

The third screen in Figure 21-3 illustrates a completed transfer of files between two calculators. During the transfer of the files, L_1 was renamed **DATA**.

If you want to terminate the transfer of files in progress, press ON on either calculator. Then press ENTER when you're confronted with the Error in Xmit error message. If you put one calculator in Receiving mode and then decide not to transfer any files to that calculator, press ON to take it out of Receiving mode or simply wait for the transfer to time out.

Transferring Files to Several Calculators

After transferring files between two calculators (as described in the preceding section), you can then use the sending and/or receiving calculator to transfer the same files to a third calculator, usually without having to reselect the files. (If the initial transfer consisted of files selected from the **All–** submenu of the **Link SEND** menu, then you will have to reselect the files.)

To transfer files to a third calculator, follow these steps:

1. **After making the initial transfer of files between the sending and receiving calculators, wait until Step 4 before pressing any keys on the calculator that will be used to transfer the files to a third calculator.**

   After the initial transfer of files is complete, the screens on the sending and receiving calculators look similar to those in Figure 21-4, only with different files. If you press any key on the calculator, other than those specified in Step 3, the files you are planning on sending to a third calculator will no longer be selected and you will have to reselect them.

2. **Link the third calculator to either the sending or receiving calculator.**

3. **On the third calculator, press** [2nd][X,T,Θ,*n*][▶][ENTER].

4. **On the other calculator, press** [2nd][X,T,Θ,*n*] **and select the same menu item that was used in the initial transfer of files. Press** [▶][ENTER] **to complete the transfer of the files to the third calculator.**

   The files from the previous transfer are still selected, provided that in the interim you made no new selection from the **Link SEND** menu (as shown in Figure 21-4). The selected files in this figure are the files that were sent to the receiving calculator in the initial transfer of files.

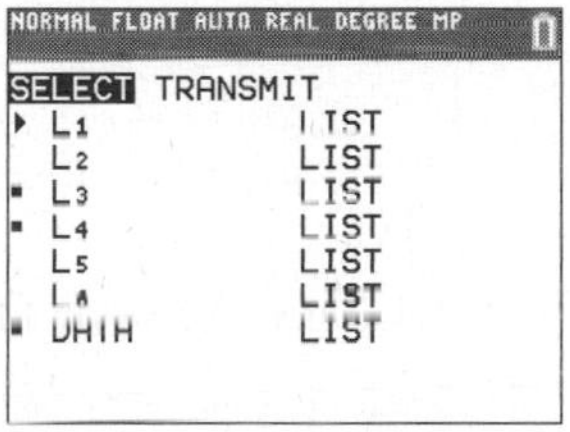

Figure 21-4:
Transferring files to a third calculator.

Chapter 22

Fun with Images

*W*hat use would it be to have a high-resolution color screen if you can't take full advantage of it? Not only is inserting color images on a graph fun, but it also serves as a way to engage students in real-world mathematics. As a special bonus, you get to see how Quick Plot & Fit Equation adds a layer of mathematics on top of an image on a graph. Of course, you can only participate in the fun if you are using the TI-84 Plus C. Keep reading and let the fun begin!

Inserting Photo Images as a Background of a Graph

An image is a digital picture that can serve as the background for your graph screen. Inserting an image is a great backdrop to practice transforming functions.

Some images have been preloaded on your calculator. To insert an image that has been preloaded, follow these steps:

1. **Press [2nd][ZOOM] to access the Format menu.**

2. **Use the [▲] key to navigate your cursor to Background.**

    When your cursor is on Background, you get a preview of the color (or image) that takes up about half of the screen.

3. **Use the ▶◀ keys to operate the menu spinner until you find the image you want.**

   You can store up to ten images on your calculator. See the first screen in Figure 22-1. Image1 through image5 are preloaded on your calculator.

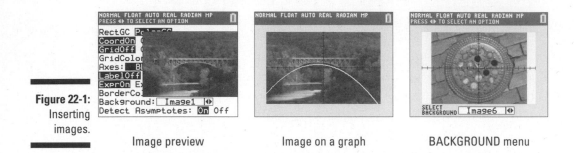

Figure 22-1:
Inserting
images.

    Image preview   Image on a graph   BACKGROUND menu

4. **Press GRAPH to see the image displayed as the background of the graph screen.**

   In the second screen in Figure 22-1, see my first attempt at transforming a parabola to fit the bridge image.

Alternatively, you can access images directly from the graph. Press GRAPH, then press 2nd PRGM ◀ and press ENTER to select BackgroundOn. These actions produce a spinner so that you can preview the images right on the graph as shown in the third screen in Figure 22-1.

Using TI Connect Software to Transfer Images

See Chapter 20 for instructions on downloading, installing, and running TI Connect software on your computer. Once you have TI Connect software up and running, use the USB computer cable that came with your calculator to connect the calculator to the computer. Click on the Device Explorer icon in the TI Connect software as shown in the first window in Figure 22-2.

The Device Explorer window can be used to drag and drop images from your computer to your calculator, or vice versa. Just open a computer documents folder and a Device Explorer window at the same time and let the dragging and dropping fun begin!

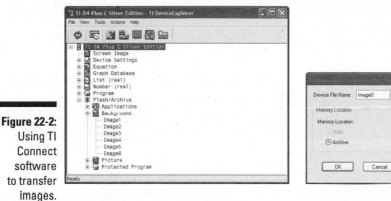

Figure 22-2:
Using TI
Connect
software
to transfer
Images.

Device Explorer window   Naming the image

You are allowed to use a GIF, TIF, PNG, JPG, or BMP file. When you drag
and drop one of these files to the Device Explorer window, TI Connect converts the file to an .8ca image file. 8ca calculator files are 83-x-133 pixels and
use 16-bit color. As soon as you drag and drop the file, a Device File Name
window opens on the computer as shown in the second window in Figure 22-2.
Select the location of the image using the drop-down menu and click OK.

Unit-to-Unit Image Transfer

You can only send images from a TI-84 Plus C calculator to another TI-84 Plus
C calculator. Images will not display on a TI-84 Plus calculator because of the
differences in screen resolution. Here are the steps to transfer images:

1. **On the receiving calculator, press** [2nd][X,T,Θ,*n*][▶][ENTER].

   You see a screen that says **Waiting**, and in its upper-right corner, a
   moving dashed line indicates that the receiving calculator is waiting to
   receive files. See the first screen in Figure 22-3.

2. **Press** [2nd][X,T,Θ,*n*] **on the sending calculator to access the Link SEND
   menu.**

   See the second screen in Figure 22-3.

3. **Press** [6] **to select Pic & Image in the Link SEND menu.**

4. **Use the** [▲][▼] **keys to move the cursor to a file you want to send, and
   press** [ENTER] **to select that file. Repeat this process until you have selected
   all the files in this list that you want to send to another calculator.**

The calculator places a small square next to the files you select, as in the third screen in Figure 22-3. In this screen, **Image2** and **Image4** are selected in the Pic & Image SELECT menu.

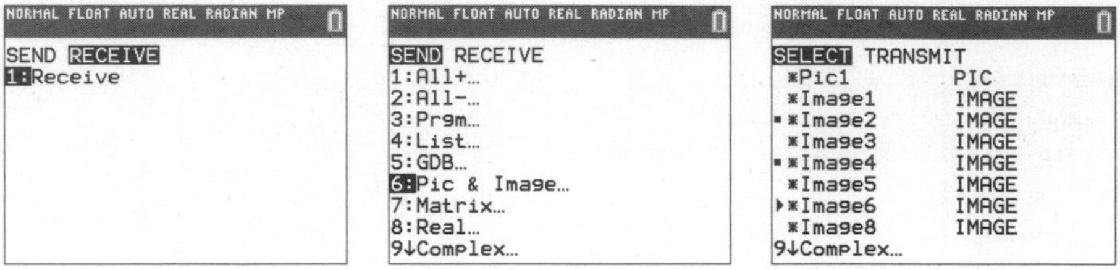

Figure 22-3:
Sending images from unit to unit.

Receiving calculator   Sending calculator   Selecting images

5. **On the sending calculator, press ▶ to access the Link TRANSMIT menu.**

   Always put the receiving calculator in Receiving mode *before* you transfer files from the sending calculator! If you forget, you will get the `ERROR: Error in Xmit` error message.

6. **On the sending calculator, press ENTER to send the files to the receiving calculator.**

Using Quick Plot & Fit Equation

Images on a graph screen have another purpose that is now a built-in tool! You can use Quick Plot & Fit Equation to quickly plot points directly on your graph and perform a regression on the points you so quickly plotted. Okay, I think I may have mentioned that this feature is quick to use, but the steps are pretty intuitive as well once you get started.

1. **Load a background image and set an appropriate graphing window**

   See the earlier section, "Inserting Photo Images as the Background of a Graph."

2. **Press STAT ▶ ▲ to locate Quick Plot & Fit–EQ.**

   See the first screen in Figure 22-4.

3. **Press ENTER to activate Quick Plot & Fit–EQ.**

4. **Use the ▶ ◀ ▲ ▼ keys to navigate your cursor, press ENTER to plot a point, and repeat until you have enough points for a regression.**

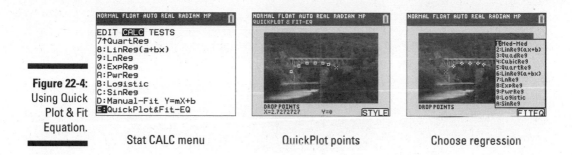

Figure 22-4:
Using Quick
Plot & Fit
Equation.

Stat CALC menu   QuickPlot points   Choose regression

See the second screen in Figure 22-4.

You need at least two points for a linear regression, at least three points for a quadratic regression, and so on.

5. Press GRAPH to activate the FITEQ on-screen prompt.

See the third screen in Figure 22-4.

6. Use the ▼ ▲ keys to navigate to the regression you want and press ENTER.

See the first screen in Figure 22-5.

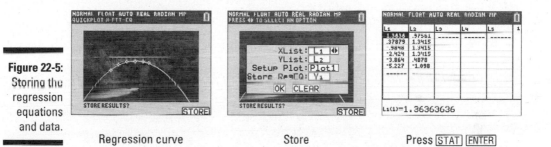

Figure 22-5:
Storing the
regression
equations
and data.

Regression curve   Store   Press STAT ENTER

7. Press GRAPH to activate the STORE on-screen prompt.

See the second screen in Figure 22-5.

8. Use the ▶ ◀ ▲ ▼ keys to navigate the menu spinners and press ENTER or OK to store.

You are storing two lists, one stat plot, and one equation. See the list in the third screen in Figure 22-5. Now that the regression equation is stored, you can trace and analyze the graph of the regression equation as much as you desire.

Chapter 23

Managing Memory

- -

In This Chapter

▶ Checking available memory

▶ Deleting and archiving to preserve memory

▶ Resetting the calculator

▶ Grouping and ungrouping programs

▶ Garbage collecting to create more usable memory

- -

First, a quick comparison: The TI-84 Plus Silver Edition has 1.5 MB of available memory. The new TI-84 Plus C Silver Edition has a whopping 3.5 MB of available memory. Why the large discrepancy? The TI-84 Plus C needs more memory so that it can store up to ten images in its archive memory. Of course, both calculators have about the same amount of random access memory (RAM).

If you don't know the difference between RAM and archive memory, you need to keep reading. If you frequently use your calculator, you are likely to run into memory issues at one point or another. This chapter helps you successfully navigate any memory issue you may come across.

Checking Available Memory

Before I explain memory in a little more detail, I need to show you how to access the available memory on your calculator. Follow these steps:

1. **Press 2nd + to access the MEMORY menu.**

   See the first screen in Figure 23-1.

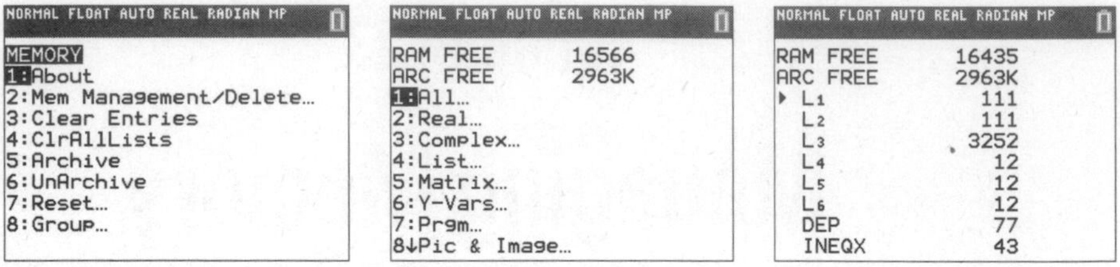

Figure 23-1:
Checking
available
RAM and
archive
memory.

NORMAL FLOAT AUTO REAL RADIAN MP		NORMAL FLOAT AUTO REAL RADIAN MP		NORMAL FLOAT AUTO REAL RADIAN MP	
MEMORY		RAM FREE	16566	RAM FREE	16435
1:About		ARC FREE	2963K	ARC FREE	2963K
2:Mem Management/Delete…		**1:**All…		▸ L₁	111
3:Clear Entries		2:Real…		L₂	111
4:ClrAllLists		3:Complex…		L₃	3252
5:Archive		4:List…		L₄	12
6:UnArchive		5:Matrix…		L₅	12
7:Reset…		6:Y-Vars…		L₆	12
8:Group…		7:Prgm…		DEP	77
		8↓Pic & Image…		INEQX	43

Memory menu   Press ②   List memory

2. Press ② to display the Memory Management/Delete menu.

See the second screen in Figure 23-1. Conveniently located at the top of the screen is a display of the amount of available random access memory (RAM FREE) and archive (ARC FREE.) You can check the amount of memory each variable type is using by selecting a menu item. I pressed ④ for list as shown in the third screen in Figure 23-1.

Deleting and Archiving to Preserve Memory

There are major differences between RAM and archive memory. RAM memory stores computations, lists, variables, data, and programs that are not archived. Archive memory stores apps, groups, pics, images, and programs or other variables that have been archived.

Here is what you need to keep in mind: Storing items in RAM memory is a risky proposition. If the battery on your calculator goes dead, you may lose all the items stored in the RAM. Or, if you accidentally drop your calculator, you may get a "RAM cleared" message on your calculator screen.

The archive memory on the TI-84 Plus C holds close to 3.5 MB of available archive space. Did you know that you can load up to 216 apps on your calculator? With all that available archive memory, there is a good chance you will never need more archive memory. If you need more archive memory, I recommend deleting some of the preloaded apps that are in a language you are not familiar with.

If you have a lot of programs on your calculator, then you may run low on available RAM. You can delete programs you don't want as long as you are sure you don't want to use them again. Or, you can archive a program. Archiving a program keeps it safe by storing the program in archive memory. The only drawback to archiving a program is that you have to unarchive the program if you want to execute it.

Follow these steps to delete, archive, or unarchive programs all using the same screen:

1. **Press** 2nd + 2 **to display the Memory Management/Delete menu.**

2. **Press** 7 **to display the Program editor screen.**

   See the first screen in Figure 23-2.

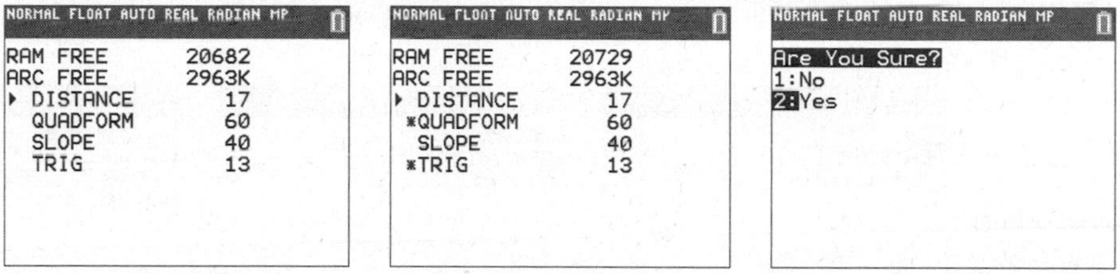

Figure 23-2:
Archiving,
unarchiving,
and deleting
programs.

Program editor	Archiving programs	Deleting programs

3. **Use the** ▲▼ **keys to select a program you want to archive and press** ENTER **to archive the program.**

   An asterisk (*) symbol appears next to programs that have been archived. See the second screen in Figure 23-2.

4. **Use the** ▲▼ **keys to select an archived program and press** ENTER **to unarchive the program.**

   Notice, the asterisk (*) symbol disappears when you unarchive a program.

5. **Use the** ▲▼ **keys to select a program to delete and press** DEL.

   See the third screen in Figure 23-2.

6. **Use the** ▲▼ **keys to select your answer to the question and press** ENTER.

   Are you sure you want to delete the program you selected? If you change your mind, select No. See the third screen in Figure 23-2.

Resetting the Calculator

There are many options when it comes to resetting your calculator. To access the RAM ARCHIVE ALL menu, press 2nd + 7. Use the ▶◀ keys to navigate the three drop-down menus. There are two choices on the RAM menu, as shown in the first screen in Figure 23-3:

✔ **Defaults:** Restores the default factory settings to all system variables, including the mode settings.

✔ **ALL RAM:** All your programs and data stored in RAM will be erased. In addition, the default factory settings are restored.

After selecting a reset option, you are given a chance to change your mind, as shown in the second screen in Figure 23-3. Sometimes a warning message is displayed that reminds you of the severity of what you are about to do if you choose to continue with the reset.

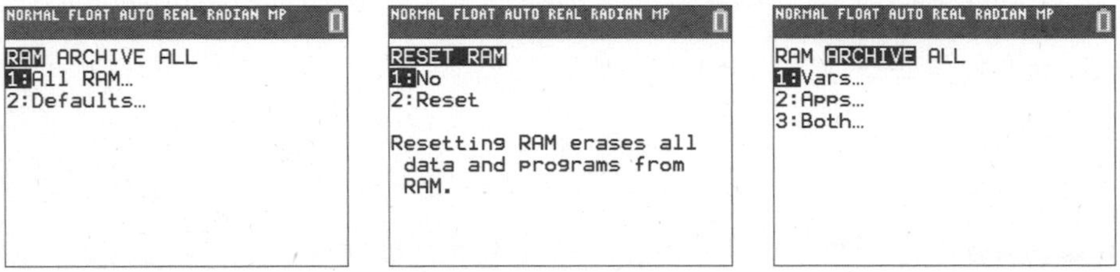

Figure 23-3:
Resetting your calculator.

RAM resets	Warning message	ARCHIVE resets

There are three choices in the ARCHIVE drop-down menu, as shown in the third screen in Figure 23-3:

✔ **Vars:** All the data stored in archive memory will be lost.

✔ **Apps:** All the apps on your calculator will be deleted.

✔ **Both:** All the data and apps will be deleted.

If you want to start from scratch, the ALL drop-down menu contains only one earth-shattering choice:

✔ **All Memory:** I think the message displayed says it all: "Resetting ALL will delete all data programs & Apps from RAM & Archive." One interesting note — the Finance app is the only app that will not be erased by executing this procedure. Some state tests require all memory be cleared before the administration of the state exam. Other states require your calculator be put in Press-to-Test mode.

Grouping and Ungrouping Programs

You can group variables residing in RAM memory and store the group in archive memory for safekeeping. Then, if the RAM on your calculator is cleared, you can ungroup the variables and they will be restored in their original state. It's a great idea to group the programs stored in the RAM of your calculator. Here are the steps to group and ungroup programs:

1. **Press 2nd+8 to access the GROUP UNGROUP menu.**

   See the first screen in Figure 23-4.

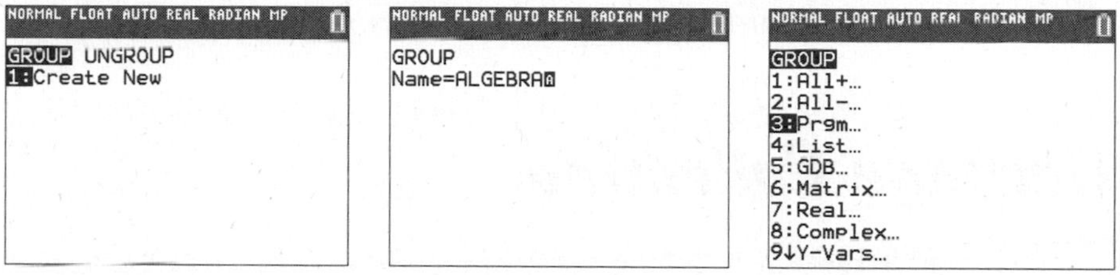

Figure 23-4: Creating a group.

GROUP UNGROUP menu   Enter group name   Select program

2. **Press ENTER to create a new group.**

3. **Enter a name for the group and press ENTER.**

   See the second and third screens in Figure 23-4.

4. **Pressing 3 enables you to select programs from a list.**

5. **Use the ▲▼ keys and press ENTER to select programs you wish to group.**

   Notice the small square next to the programs you have selected. See the first screen in Figure 23-5.

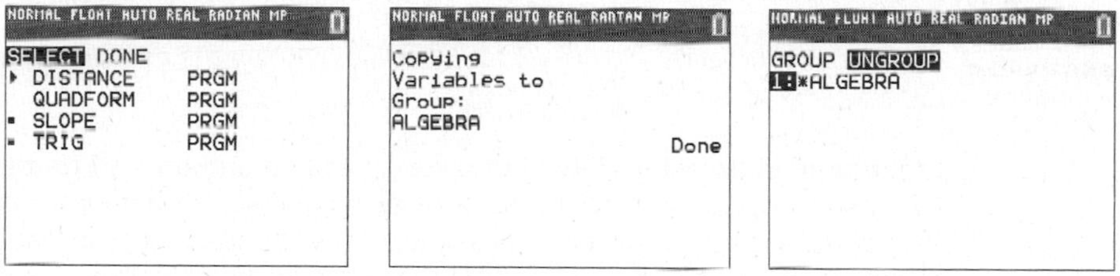

Figure 23-5: Creating a group and then ungrouping.

Select programs   Select DONE   Ungrouping list

6. **Press ▶ENTER to select DONE.**

   You have successfully created a group of programs. If the RAM on your calculator is cleared, you now have a backup of your programs. See the second screen in Figure 23-5.

7. **To ungroup, press 2nd + 8 ▶ to see a list of groups.**

   See the third screen in Figure 23-5. Ungrouping will not delete the original group. The group remains in archived memory until you decide to delete the group.

8. **Use the ▲▼ keys and press ENTER to select the group you wish to ungroup.**

   If you ungroup a program that already exists in the RAM, you will be given the option to rename or overwrite the program.

Garbage Collecting

Sometimes, you get the ERROR: ARCHIVE FULL error message in spite of the fact that you seem to have plenty of memory available. What gives? If you have just made major changes, like deleting apps, your calculator is not able to use all the available memory until it reorganizes the files. This reorganization has a funny name — garbage collecting. To create more usable memory space, a "Garbage Collect?" prompt displays, as shown in the first screen in Figure 23-6.

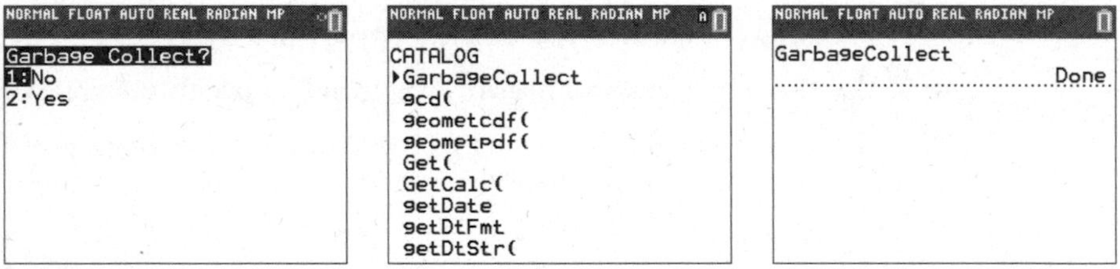

Figure 23-6: Garbage collecting.

Garbage Collect?　　　　　Catalog　　　　　Forcing Garbage Collect

I recommend pressing 2 for Yes, as long as you understand that it could take as long as 20 minutes to execute garbage collecting on your calculator. You should expect a process message, defragmenting as your calculator reorganizes its files. You also don't have to wait until your calculator forces you to action. Why not be proactive? Press 2nd 0 to access the Catalog and press TAN to jump to the commands beginning with the letter G, as shown in the second screen in Figure 23-6. Press ENTER to insert the GarbageCollect command on the Home, screen and press ENTER again to begin the potentially long process. The final result is illustrated in the third screen in Figure 23-6.

Part VI
The Part of Tens

Enjoy a list of ten common keystroke sequences online at www.dummies.com/
extras/ti84plus.

In this part . . .

- ✔ Learn to perform the ten most essential calculator tasks that you may come across in a math or science classroom.

- ✔ Review the ten most common mistakes so that you can avoid making them.

- ✔ Look at a directory of the ten most common error messages with explanations of the cause of the problem.

Chapter 24

Ten Essential Skills

● ●

In This Chapter
▶ Reviewing six essential calculator skills for use on the Home screen
▶ Reviewing four essential graphing skills

● ●

1n this chapter, you find a brief review of some of the most important and basic calculator skills that are explained in the book. If you need more detailed instructions (with accompanying screenshots), the chapter location is referenced for each of the ten skills.

Copying and Pasting

Save time by copying and pasting expressions on the Home screen. Press the ▲ key to scroll through your previous calculations. When a previous entry or answer is highlighted, press ENTER to paste it into your current entry line. After you have pasted the expression into the current entry line, you can edit the expression as much as you like. For a more detailed description, see Chapter 1.

Converting a Decimal to a Fraction

Converting a decimal to a fraction is quick and easy. On the Home screen, press MATH ENTER ENTER. If you want to convert a fraction to a decimal, press MATH ▼ ENTER or include a decimal point in your calculations. See more in Chapter 3.

Changing the Mode

Before you get too far, make sure you can adjust the mode settings on your calculator. The Status bar at the top of the screen indicates some of the most popular mode settings. To see the rest of the mode settings, press MODE. Use the ▶◀▲▼ keys to highlight the setting you want, and press ENTER to select the highlighted setting. Many chapters in this book contain detailed instructions on setting the mode to accomplish desired tasks.

Accessing Hidden Menus

Did you know that there are four hidden shortcut menus on your calculator? The four menus are: FRAC (Fraction menu), FUNC (Function menu), MTRX (Matrix menu), and YVAR (Y-variables menu). To access the FRAC hidden menu, press ALPHA Y=. After pressing ALPHA, the keys at the top of your keypad become soft keys that activate on-screen menus. Press ALPHA Y= to access the fraction menu and press ALPHA WINDOW to find the math templates. You can access the MTRX shortcut menu only by pressing ALPHA ZOOM (it is not found in standard menus). Finally, press ALPHA TRACE to locate the Y-Var shortcut menu. Find out more details in Chapter 3.

Entering Imaginary Numbers

You can enter an expression that includes the imaginary number, *i*, by pressing 2nd .. Your calculator automatically simplifies expressions containing imaginary numbers.

Complex numbers cannot be used with the n/d fraction template. Instead, enter complex numbers as fractions using parentheses and the ÷ key.

If you want to find out more details about using imaginary numbers, see Chapter 5.

Storing a variable

The letters STO may look like texting language, but the STO▶ key on a calculator is a handy feature. If you plan to use the same number many times when evaluating arithmetic expressions, consider storing that number in a variable. To do so, follow these steps:

1. **Enter the number you want to store in a variable.**

2. **Press** STO▸.

3. **Press** ALPHA **and press the key corresponding to the letter of the variable in which you want to store the number.**

4. **Press** ENTER **to store the value.**

After you have stored a number in a variable, you can type the variable to insert that number into an arithmetic expression. To do so, press ALPHA, and press the key corresponding to the letter of the variable in which the number is stored. See Chapter 2 for a more detailed description of storing variables.

Graphing a Function

Entering and graphing functions are on two different screens. Here are the steps to graph a function.

1. **Press** Y= **to access the Y= editor.**

2. **Enter your function.**

   Press X,T,Θ,n to insert an X in the function.

3. **Press** ZOOM 6 **to graph the function in a standard viewing window.**

Of course, there are many tools that you can use to customize your graph. Chapter 9 explains how to adjust the window, change the color, set the format of your graph, and more.

Finding the Intersection Point

Given two functions, can you find the intersection point? To do so, press Y= and enter the functions in Y_1 and Y_2. Press ZOOM 6 to graph both functions. If you can see the point of intersection on the screen, press 2nd TRACE 5 ENTER ENTER ENTER to find the point of intersection of the graph. If you can't see the point of intersection or have more than one point of intersection, then see Chapter 11 for a more complete description of finding intersection points.

Graphing a Scatter Plot

Graphing a scatter plot is a multistep process:

1. **Press [STAT][ENTER] and enter data in lists L₁ and L₂.**
2. **Press [2nd][Y=][ENTER] and configure the Plot1 editor.**

   Highlight ON and press [ENTER]. Then highlight ⌐∵ and press [ENTER] to configure the plot editor for a scatter plot.
3. **Press [ZOOM][9] to graph the scatter plot in a ZoomStat window.**

To change the color or further customize your scatter plot, see Chapter 18.

Performing a Linear Regression

Many different regressions can be performed on data, though the most common regression is a linear regression. To perform a linear regression, follow these steps:

1. **Press [STAT][ENTER] and enter data in lists L₁ and L₂.**
2. **Press [STAT][▶][4] to start the LinReg(ax+b) Wizard.**
3. **Configure the wizard and press [ENTER] repeatedly until the regression results appear on the screen.**

For a more detailed description of performing regressions, see Chapter 18.

Chapter 25

Ten Common Errors

*E*ven the best calculating machine is only as good as its input. This chapter identifies ten common errors made when using the calculator. Wouldn't it be great to avoid some of the common errors that normally plague students who are using calculators?

Using ⊟ Instead of ⟮-⟯ to Indicate That a Number Is Negative

If you press ⊟ Instead of ⟮-⟯ at the beginning of an entry, the calculator assumes you want to subtract what comes after the minus sign from the previous answer. If you use ⊟ Instead of ⟮-⟯ in the interior of an expression to denote a negative number, the calculator responds with the ERROR: SYNTAX error message.

Indicating the Order of Operations Incorrectly by Using Parentheses

When evaluating expressions, the order of operations is crucial. To the calculator, for example, -3^2 equals -9. This may come as quite a surprise to someone expecting to square -3, where $(-3)^2$ equals 9. The calculator first performs the operation in parentheses, then it squares the number, and if there is a negative outside the parentheses, it first performs the squaring and then the operation of negating a number. Unless you're careful, this won't provide the answer you're looking for. To guard against this error, you may want to review the detailed list of the order in which the calculator performs operations (see Chapter 2).

Improperly Entering the Argument for Menu Functions

If an argument is improperly entered, a menu function won't work. A prime example is the **fMin** function housed in the Math MATH menu. Do you remember what to place after this function so that you can use it? If you don't, you get the ERROR: ARGUMENT error message.

To avoid this error, you can use the Catalog Help feature to see the syntax of the function you would like to use. Press MATH and use the ▼ key to move your cursor to the **fMin** function as shown in the first screen in Figure 25-1. Press ➕ to access the Catalog Help feature as illustrated in the second screen in Figure 25-1.

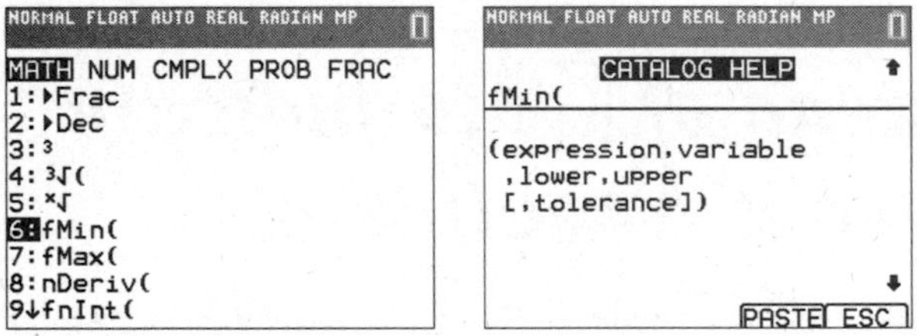

Figure 25-1: Accessing the built-in Catalog Help feature.

MATH menu   Press ➕

Accidentally Deleting a List

If your cursor is in the column heading and you press DEL, the list disappears from view. List L_2 isn't displayed in the first screen in Figure 25-2. Don't worry! You can recover the list by using the SetUpEditor command. Press STAT 5 ENTER, as shown in the second screen in Figure 25-2. List L_2 is now restored in the List editor. Press [STAT][ENTER] to see the lists, as shown in the third screen in Figure 25-2.

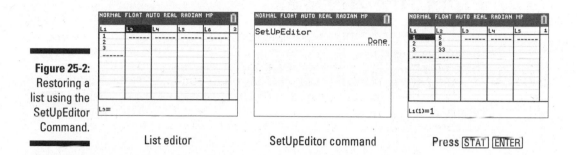

Figure 25-2:
Restoring a
list using the
SetUpEditor
Command.

List editor   SetUpEditor command   Press [STAT] [ENTER]

Entering an Angle in Degrees in Radian Mode

To change the mode, press [MODE], move your cursor to DEGREE and press [ENTER]

Alternatively, you *can* enter an angle in degrees when you are in Radian mode, but you have to let the calculator know that you're overriding the Angle mode by placing a degree symbol after your entry. To insert a degree symbol, press [2nd][APPS][ENTER]. See Chapter 7 for a more detailed description of entering angles in your calculator.

Graphing Trigonometric Functions in Degree Mode

This, too, is a mistake unless you do it just right: In the Window editor, you have to set the limits for the x-axis as $-360 \leq x \leq 360$. Pressing [ZOOM][7] or [ZOOM][0] to have the calculator graph the function using the **ZTrig** or **ZoomFit** command produces similar results. But this works when you're graphing pure trig functions such as $y = \sin x$. If you're graphing something like $y = \sin x + x$, life is a lot easier if you graph it in Radian mode.

Graphing Functions When Stat Plots Are Active

If you get the ERROR: DIMENSION MISMATCH error message when you graph a function, this is most likely caused by a stat plot that the calculator is trying to graph along with your function. You can turn off the stat plot on the Y= editor screen. Press [Y=], and see if any of the stat plots are highlighted at the top of the Y= editor screen. Stat Plot1 and Stat Plot3 are highlighted in the first screen in Figure 25-3. Move your cursor over the highlighted stat plots and press [ENTER] to turn off the stat plots, as illustrated in the second screen in Figure 25-3.

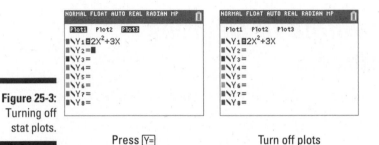

Figure 25-3:
Turning off
stat plots.

Press [Y=]   Turn off plots

Inadequately Setting the Display Contrast

If your screen is too light or too dark to read the calculations, you can fix it easily. To adjust the contrast settings to your liking, repeatedly press [2nd][▲] to darken the screen or press [2nd][▼] to lighten the screen.

Setting the Window Inappropriately for Graphing

If you get the ERROR: WINDOW RANGE error message when graphing functions, this is most likely caused by setting **Xmin** $\geq$ **Xmax** or by setting **Ymin** $\geq$ **Ymax** in the Window editor. Setting the Window editor is explained in Chapter 9.

Accidentally Deactivating a Function

This can be one of the most frustrating mistakes you can make. You have to be paying attention to notice that a function has been deactivated. See the first screen in Figure 25-4. Notice the equal sign next to Y_1 isn't highlighted. This means the function has been deactivated.

To activate a function that has been deactivated, move your cursor to the equal sign in the Y= editor and press ENTER.

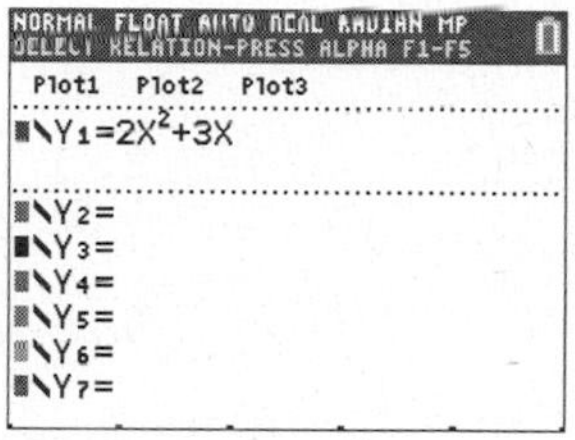

Figure 25-4:
Activating a
deactivated
function.

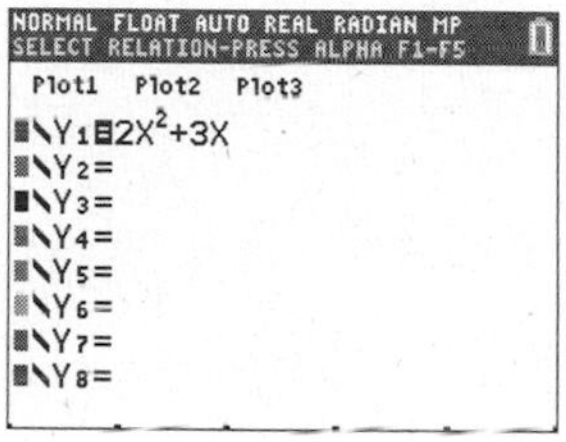

Y₁ is deactivated   Y₁ is activated

Chapter 26

Ten Common Error Messages

*T*his chapter gives a list of ten common error messages the calculator may give you. When you get an error, pay close attention to the error message screen. Your calculator displays a descriptive note on the error screen that helps you determine the cause of the error.

ARGUMENT

You usually get this message when you are using a function housed in one of the menus on the calculator. This message indicates that you have not properly defined the argument needed to use the function.

To avoid this error, take advantage of the Catalog Help feature. Use the ▶◀▲▼ keys to navigate your cursor to the function you want and press ⊞ to view the syntax of the function you want.

BAD GUESS

This message indicates that the guess you've given to the calculator isn't within the range of numbers that you specified. This is one of those times when the calculator asks you to guess the solution. One example is when you're finding the maximum value or the zero of a function within a specified range (see Chapter 11). Another is when you're finding the solution to an equation where that solution is contained in a specified range (see Chapter 4).

One other time that you can get this message is when the function is undefined at (or near) the value of your guess.

DIMENSION MISMATCH

You usually get this message when you attempt to add, subtract, or multiply matrices that don't have compatible dimensions.

You also get this error if you try to graph a scatter plot of data lists that don't have the same dimensions. In other words, the number of elements in L_1 and L_2 are not the same.

DOMAIN

You usually get this message when you're using a function housed on a menu of the calculator. If that function is, for example, expecting you to enter a number in a specified range, you get this error message if that number isn't in the specified range.

When using the Finance app, discussed in Chapter 19, you get this message when you don't use the correct sign for cash flow.

INVALID

This is the catchall error message. Basically, it means that you did something wrong when defining something (for example, you used function Y_3 in the definition of function Y_2, but forgot to define function Y_3).

INVALID DIMENSION

You get this invalid-dimension message if (for example) you attempt to raise a nonsquare matrix to a power or enter a decimal for an argument of a function when the calculator is expecting an integer.

You can also get this error if you inadvertently leave the stat plots on when you are trying to graph a function. Turn the stat plots off by pressing [2nd][Y=][4].

NO SIGN CHANGE

When you're using the Equation Solver (detailed in Chapter 4), you get this message when the equation has no real solutions in your specified range.

SINGULAR MATRIX

You get this message when you try to find the inverse of a matrix whose determinant is zero.

SYNTAX

This is another catchall error message. It usually means you have a typo somewhere or you have done something the calculator wasn't expecting.

WINDOW RANGE

This, of course, means that the Window is improperly set. This problem is usually (but not always) caused by improperly setting **Xmin** ≥ **Xmax** or **Ymin** ≥ **Ymax** in the Window editor. For a look at the proper way to set the Window for functions, check out the explanations in Chapter 9.

Troubleshooting a Calculator that is Not Functioning Properly

If your TI-84 Plus C will not turn on or is not behaving properly, you might need to resort to more serious measures. Try these steps in order until you are able to fix your malfunctioning calculator:

1. **Press the reset button the back of the calculator to reboot.**

   This will clear the RAM on your calculator.

2. **Hold down the Reset button and** DEL**, then release only the Reset button.**

   This sequence causes a "Waiting . . ." message to display on the Home screen. You must reinstall the OS (see Chapter 20.)

3. **Use a small screwdriver to remove the lithium battery for five minutes, then reinsert the battery and turn the calculator on.**

Appendix A

Creating Calculator Programs

* * *

In This Chapter

▶ Creating, saving, editing, and deleting calculator programs

▶ Editing a program on the calculator

▶ Executing programs on the calculator

▶ Deleting a calculator program

▶ Using a computer to enter a calculator program

* * *

    *T*he programming language used by the calculator is similar to the Basic programming language. It uses the standard commands (such as the "If ..., then ..., else ..." command) that are familiar to anyone who has ever written a program. And, of course, it also makes use of commands that are unique to the calculator (such as **ClrHome**, which clears the Home screen). This appendix explains the basics of creating a calculator program. Appendix B discusses programming commands that are unique to the calculator, and Appendix C describes the Basic programming commands used by the calculator.

Creating and saving a program on the calculator

These are the basic steps for creating a program on the calculator:

1. **Press** PRGM ◀ ENTER **to create a new program using the Program editor.**

   This is illustrated in the first screen in Figure A-1.

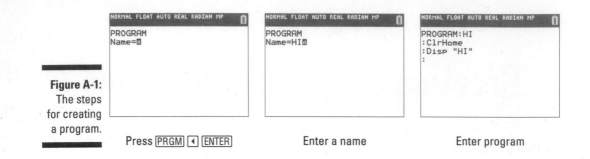

Figure A-1:
The steps
for creating
a program.

Press PRGM ◄ ENTER   Enter a name   Enter program

2. **Give your program a name and then press** ENTER.

   The name of your program can consist of one to eight characters that must be letters, numbers, or the Greek letter θ. The first character in the name must be a letter or θ, as in the second screen in Figure A-1.

   The 🔒 after **Name** = indicates that the calculator is in Alpha mode. In this mode, when you press a key you enter the green letter above that key. To enter a number, press ALPHA to take the calculator out of Alpha mode and then enter the number. To enter a letter after entering a number, you must press ALPHA to put the calculator back in Alpha mode. Press 2nd ALPHA to place your calculator in Alpha-lock mode which allows you to enter multiple letters without having to press ALPHA between them.

   When you press ENTER after naming your program, the calculator puts you in the Program editor, as in the third picture in Figure A-1. The program appearing in this screen is entered in the next step.

3. **Enter your program in the Program editor.**

   Your program consists of a series of commands, each of which must be preceded by a colon, as shown in the third screen in Figure A-1. After entering a command, press ENTER so the calculator supplies the colon preceding the next command you enter. When you finish writing your final command, press ENTER and ignore the colon that is waiting for a command to be entered.

   An example of entering a program appears in the third screen in Figure A-1. The program in this screen writes HI on the Home screen. The **Disp** command is entered by pressing PRGM ► 3 ENTER.

4. **Press** 2nd MODE **when you're finished writing your program.**

   This saves your program in the memory of the calculator and returns you to the Home screen. The name under which the program is stored in the calculator is the same name you gave the program in Step 2.

Editing a program on the calculator

To edit a program stored on the calculator, follow these steps:

1. **Press PRGM▶ and press the number of the program or use the ▲▼ keys to highlight the program you want to edit.**

2. **Edit the program.**

   Pressing CLEAR deletes the line containing the cursor.

3. **Press 2nd MODE to save the program and return to the Home screen.**

Executing a Calculator Program

After creating your program and saving it on the calculator, you can run the program on the calculator by performing the following steps:

1. **Press PRGM to enter the Program Execute menu, and use the ▼ key to move the indicator to your program.**

   This is illustrated in the first screen in Figure A-2.

2. **Press ENTER to place the program on the Home screen.**

   This is illustrated in the second screen in Figure A-2.

3. **Press ENTER to execute the program.**

   This operation is shown in the third screen in Figure A-2. When the calculator is finished executing the program, it writes Done on the Home screen.

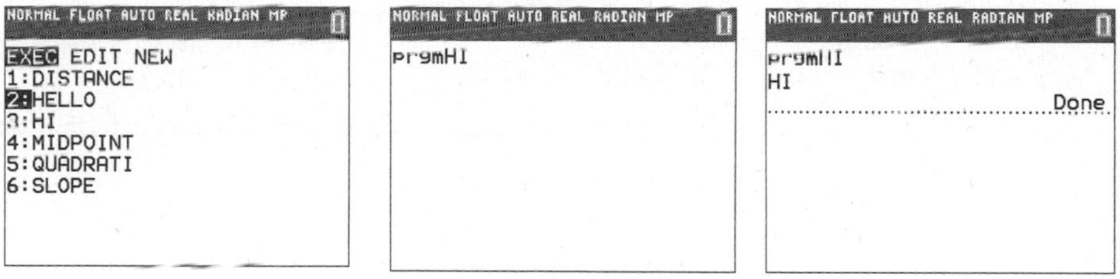

Figure A-2: Executing a program.

Program list   Select program   Press ENTER

Deleting a Program from the Calculator

To delete a program from the calculator:

1. **Press 2nd+ to access the Memory menu.**

2. **Press 2 to access the Mem Mgt/Del menu.**

3. **Press 7 to access the Program files stored in the calculator.**

4. **If necessary, repeatedly press the ▾ key to move the indicator to the program you want to delete.**

5. **Press DEL to delete the program.**

   You are asked whether you really want to delete this program. Press 2 if you want it deleted or press 1 if you've changed your mind.

6. **Press 2nd MODE to exit this menu and return to the Home screen.**

Calculator Programming on a Computer

It is much easier to create a calculator program on a computer than it is to create one on the calculator. Unfortunately, the software needed to perform this task is not available for all types of computers. On a Mac computer, TI Connect software has a built-in programming interface that can be used to program a calculator on a computer. On a PC computer, TI-Graph Link software contains a programming editor that can be used to enter a calculator program on a computer. Unfortunately, at the time of writing this book, TI-Graph Link software is not compatible with Windows 7 or Windows 8 computers.

If the software is compatible with your computer, you can create a program on the computer and then transfer it to your calculator (or vice versa). To do so, you need to download and install TI Connect software and a USB computer cable to connect your calculator to your computer. The software is free; the USB computer cable, if it didn't come with your calculator, can be purchased at the Texas Instruments online store at www.education.ti.com.

Appendix B

Controlling Program Input and Output

• •

In This Chapter

▶ Entering program input and output commands

▶ Using input commands (Input, Prompt)

▶ Using output commands (Disp, Output)

▶ Using a program to change the color and graph style of a function

▶ Using a program to change the color of text

• •

*P*rogram input is information that the program requests from the program user. *Program output* is information passed from the program back to the program user. This chapter explains how to get a program to shuttle information back and forth between the program and the program user.

The Program I/O menu, which houses the input and output commands, is available only when you're using the Program editor to create a new program or to edit an existing program. A screen of the Program I/O menu appears in Figure B-1. Creating and editing programs are explained in Appendix A.

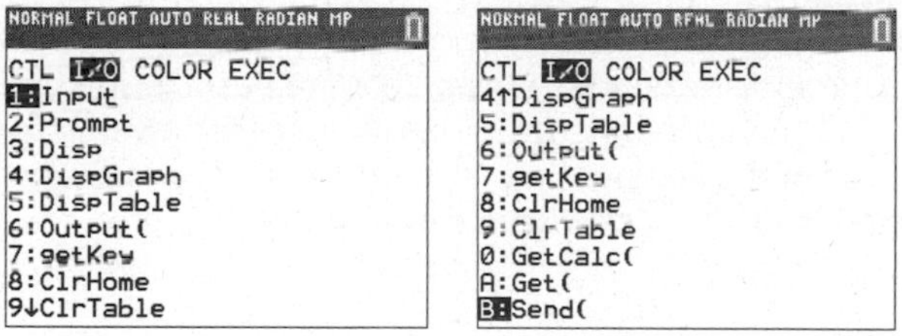

Figure B-1:
The Program I/O menu.

```
NORMAL FLOAT AUTO REAL RADIAN MP
CTL I/O COLOR EXEC
1:Input
2:Prompt
3:Disp
4:DispGraph
5:DispTable
6:Output(
7:getKey
8:ClrHome
9↓ClrTable
```

```
NORMAL FLOAT AUTO REAL RADIAN MP
CTL I/O COLOR EXEC
4↑DispGraph
5:DispTable
6:Output(
7:getKey
8:ClrHome
9:ClrTable
0:GetCalc(
A:Get(
B:Send(
```

Using Input Commands

The **Input** and **Prompt** commands are used in a program to solicit information from the program user. The **Input** command asks the user for the value of only one variable and enables the program to briefly describe the variable it is requesting. The **Prompt** command asks the user for the value of one or more variables, but doesn't allow for a description of the variable other than its name.

Using the Input command

The syntax for using the **Input** command to request the program user to assign a value to a *variable* is **Input** *"text",variable*. The *text*, which must be in quotes, offers the program user a description of what is being requested by this command. The *text* and the *variable* must be separated by a comma. Note that there is no space between the comma and the *variable*, as in the first screen in Figure B-2.

Press ALPHA + to insert quotation marks.

When the program is executed, the program displays the *text* on the Home screen and waits for the program user to enter a number and press ENTER. This is illustrated at the top of the third screen in Figure B-2. The number entered by the user is then stored in the *variable* specified by the **Input** command.

Each line of the TI-84 Plus C Home screen can accommodate a maximum of 26 characters. Up to ten lines can display at one time on the Home screen. Sometimes, this isn't enough space for the **Input** command to display the *text* and for the program user to enter the value of the *variable*. If this is the case, you may want to precede the **Input** command with a **Disp** command describing the value that the user must enter. When you do so, the syntax for the **Input** command is simply **Input** *variable*, as in the second screen in Figure B-2. When the program is executed, it displays the description given in the **Disp** command, and then prompts the program user for a value by displaying a question mark, as in the second half of the third screen in Figure B-2. Using the **Disp** command is discussed later in this chapter.

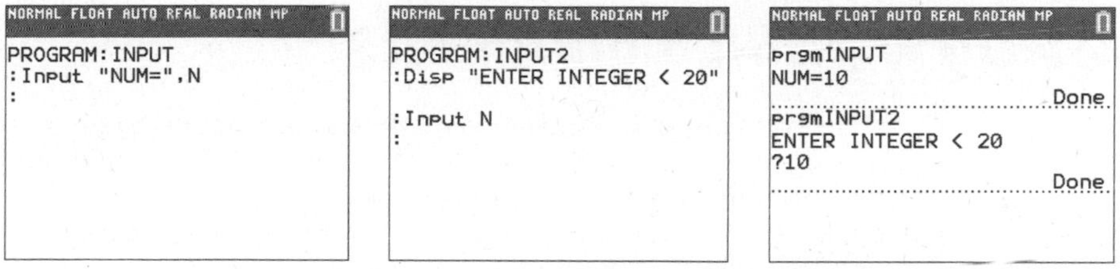

NORMAL FLOAT AUTO REAL RADIAN MP

```
PROGRAM:INPUT
:Input "NUM=",N
:
```

NORMAL FLOAT AUTO REAL RADIAN MP

```
PROGRAM:INPUT2
:Disp "ENTER INTEGER < 20"
:
:Input N
:
```

NORMAL FLOAT AUTO REAL RADIAN MP

```
prgmINPUT
NUM=10
                        Done
prgmINPUT2
ENTER INTEGER < 20
?10
                        Done
```

Figure B-2:
Using the
Input
command.

Input program   Input2 Program   Result

Using the Prompt command

The **Prompt** command asks the program user to assign values to one or more *variables*. The syntax for using the **Prompt** command is **Prompt** *variable1,variable2,...,variable n*. Commas separate the *variables* and there is no space between the comma and the next *variable*, as in the first screen in Figure B-3.

When the program is executed, the program displays the first *variable* followed by an equal sign and a question mark and waits for the program user to enter a number. It then does the same for the next *variable*, and so on, as in the second screen in Figure B-3. The numbers entered by the user are then stored in the *variable* specified by the **Prompt** command.

The Window variables **Xmin**, **Xmax**, **Ymin**, and **Ymax** are housed in the Variables Window menu. To access this menu, press [VARS][1].

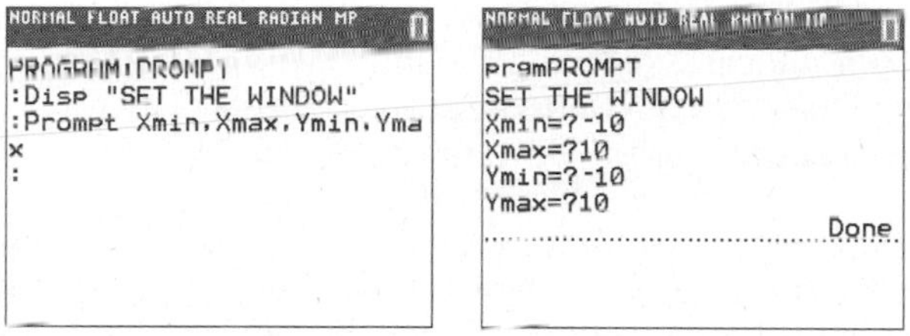

NORMAL FLOAT AUTO REAL RADIAN MP

```
PROGRAM:PROMPT
:Disp "SET THE WINDOW"
:Prompt Xmin,Xmax,Ymin,Yma
x
:
```

NORMAL FLOAT AUTO REAL RADIAN MP

```
prgmPROMPT
SET THE WINDOW
Xmin=?-10
Xmax=?10
Ymin=?-10
Ymax=?10
                        Done
```

Figure B-3:
Using the
Prompt
command.

Prompt Program   Result

Using Output Commands

The **Disp** and **Output** commands are used by a program to display text messages and values. The **Disp** command is capable of displaying more than one piece of information, and the **Output** command enables the program to place text or a value, but not both, at a predetermined location on the Home screen.

Using a program to write text

The **Disp** and **Output** commands, which are explained in the next two sections, are used to get a program to display text on the Home screen. Because each line of the Home screen can accommodate up to 26 characters, the wise programmer will limit all text items to no more than 26 characters. A space counts as one character.

The first screen in Figure B-4 shows an example of a program that displays the text "PRESS THE ENTER KEY TO CONTINUE" in two ways. The first **Disp** command displays the whole text, in spite of the fact that it contains more than 26 characters. The **Disp** command followed by an empty *text item* can be used to make a program skip a line on the Home screen. The next two **Disp** commands break the text into two parts, each of which contains fewer than 26 characters.

The output of the program in the first screen in Figure B-4 is shown in the second screen in Figure B-4. The ellipsis at the end of the second line in this screen indicates that the calculator could not display the whole line. (The calculator does not understand "wrap around.") And worse than that, you can't use ⬅➡⬆⬇ to see what comes after that ellipsis. The remaining lines of this screen illustrate the solution to this problem.

When programming the calculator to output text, limit all text items to 26 characters. A space counts as one character. If necessary, break the text into two or more text items that are consecutively displayed.

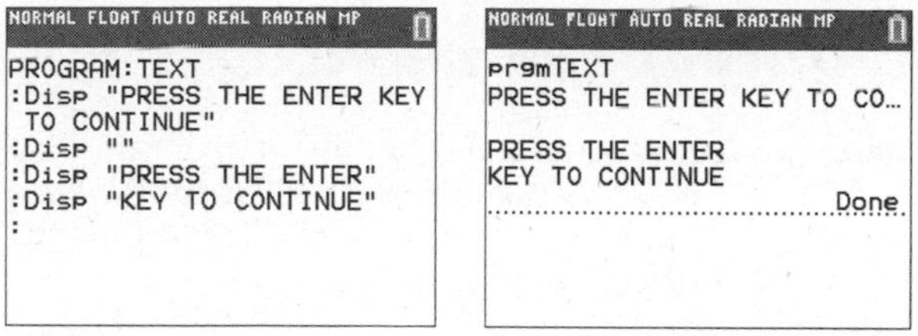

NORMAL FLOAT AUTO REAL RADIAN MP

```
PROGRAM:TEXT
:Disp "PRESS THE ENTER KEY
 TO CONTINUE"
:Disp ""
:Disp "PRESS THE ENTER"
:Disp "KEY TO CONTINUE"
:
```

NORMAL FLOAT AUTO REAL RADIAN MP

```
prgmTEXT
PRESS THE ENTER KEY TO CO...

PRESS THE ENTER
KEY TO CONTINUE
.............................Done.
```

Figure B-4:
Limiting text
items to 26
characters.

Text Program

Result

Using the Disp command

The syntax for using the **Disp** command to have a program display *text* and *values* is: **Disp** *item*1,*item*2,...,*item n* where *item* is either *text* or a *value*. The *items* appearing after this command are separated by commas with no spaces inserted after each comma. *Text items* must be in quotes, and *value items* can be arithmetic expressions, as in the last two lines of the first screen in Figure B-5.

When a program executes a **Disp** command, it places each *item* following the command on a separate line; *text items* are left justified and *value items* are right justified, as in the second screen in Figure B-5.

The Home screen, where program output is displayed, can accommodate up to ten lines. If the **Disp** command in your program is going to result in more than ten lines, consider breaking it into several **Disp** commands separated by the **Pause** command. The **Pause** command is explained in Appendix C.

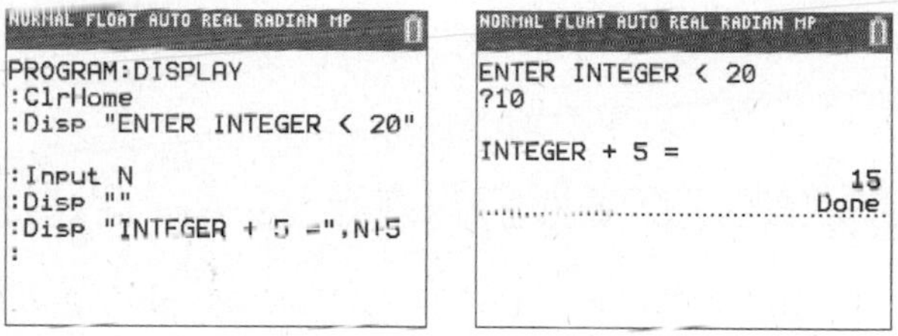

NORMAL FLOAT AUTO REAL RADIAN MP

```
PROGRAM:DISPLAY
:ClrHome
:Disp "ENTER INTEGER < 20"

:Input N
:Disp ""
:Disp "INTEGER + 5 =",N+5
:
```

NORMAL FLOAT AUTO REAL RADIAN MP

```
ENTER INTEGER < 20
?10

INTEGER + 5 =
                          15
.............................Done.
```

Figure B-5:
Using
the Disp
command.

Display Program

Result

Using the Output command

The syntax for using the **Output** command to have a program display *text* or a *value* at a specified location on the Home screen is: **Output(***line,column,item***)**. The calculator supplies the first parenthesis; you must supply the last parenthesis. There are no spaces inserted after the commas.

The Home screen contains 10 lines and 26 columns. The *item* displayed by this command can be a *text item* or a *value item. Text items* must be in quotes, and a *value item* can be an arithmetic expression, as in the last four lines of the first screen in Figure B-6. The program output resulting from executing this program is illustrated in the second screen in Figure B-6. If you look closely at the screenshot, you may notice that I did not close the parenthesis in the last line of the program: Output(6,9,N+5. Closing the parenthesis at the end of a line will have no bearing on the execution of the program. This is a convention that most programmers will use to save a tiny bit of space in the RAM of the calculator.

The Output command can be used to make text "wrap" to the next line, as shown in the second screen in Figure B-6.

```
NORMAL FLOAT AUTO REAL RADIAN MP
PROGRAM:OUTPUT
:ClrHome
:Disp "ENTER INTEGER < 20
"
:Input N
:Output(5,1,"AN INTEGER PL
UST FIVE IS EQUAL TO "
:Output(6,9,N+5
:
```

```
NORMAL FLOAT AUTO REAL RADIAN MP
ENTER INTEGER < 20
?10
                            Done
·······························
AN INTEGER PLUST FIVE IS E
QUAL TO 15
```

Figure B-6: Using the Output command.

Output Program   Result

Using a Program to Display a Graph

In this program, the **PlotsOff** command turns off all stat plots and the **FnOff** command turns off all functions in the Y= editor. The **FnOn 1,2** command turns on the first two functions in the Y= editor so that only these two functions are graphed. In order to store equations in Y_1 and Y_2, you must first put quotes around the expression you would like to graph, as shown in the first screen in Figure B-7.

The **ZStandard** command tells the calculator to graph these two functions in the standard viewing window where $-10 \le x \le 10$ and $-10 \le y \le 10$, as shown in the second screen in Figure B-7.

Commands such as **PlotsOff**, **FnOff**, and **ZStandard** can be entered in your program from the Catalog menu.

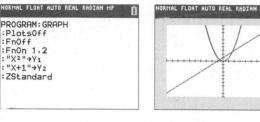

Figure B-7: Using the ZStandard command to display a graph.

Graph Program   Result

Changing the Color and Graph Style of a Function

In this program, the **GraphStyle(***function#,graph style#***)** command is used to change the attributes of a function. Entering a function number of **2** changes the graph style of function Y_2. There are eight different graph styles: 1=thin (⌐), 2=thick (▜), 3=above (▜), 4=below (▟), 5=path (◁), 6=animate (◁), 7=dot-thick (⁚), 8=dot-thin (⁚).

The **GraphColor(***function#,color#***)** command changes the color of a function. There are 15 colors to choose from: 10=Blue, 11=Red, 12=Black, 13=Magenta, 14=Green, 15=Orange, 16=Brown, 17=Navy, 18=LtBlue, 19=Yellow, 20=White, 21=LtGray, 22=MedGray, 23=Gray, and 24=DarkGray. When using the Graph Color command, you may enter its corresponding number or press VARS ◀ to access the Vars COLOR menu and make a color selection. Of course, the Graph Style and Graph Color commands should be used prior to graphing the function (see Figure B-8).

It is OK to leave off the right parenthesis at the end of a command in a program, because the program will take up less RAM on the calculator.

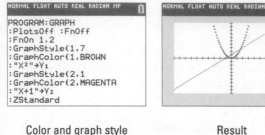

Figure B-8:
Changing
the color
and graph
style of a
graph.

Color and graph style   Result

Changing the Color of Text on a Graph

In this program, the **TextColor(***color#***)** command can be used to set the color of the text prior to using the Text command. Use the color number (see the list in the preceding section) or press VARS ◄ to access the Vars COLOR menu and make a color selection, to insert the argument of the TextColor command.

The Graph area contains 148 pixels in horizontal rows, and 256 pixels in vertical columns. The **Text(***row, column, text***)** command places text on a graph. Just because the text begins showing on the screen doesn't mean that it will fit on the screen, as shown in the second screen in Figure B-9.

In order to space from one line of text to the next, I think that 15 pixels of vertical spacing should be used to make sure text doesn't overlap.

Figure B-9:
Changing
the color
of text and
the result.

Changing text color   Result

Housekeeping Issues

Because programs display their output on the Home screen, it is a good idea to have your program clear the Home screen before the output is displayed. This is done by inserting the **ClrHome** I/O command in the program before the commands used to display the output, as in the program in the first screen in Figure B-9.

When the **Disp** I/O command is used to display the program output, it isn't necessary to clear the Home screen at the end of the program. After the program is executed, the calculator uses the next available line on the Home screen to evaluate any arithmetic expressions or to execute any commands you enter. However, you may end up typing over the top of text from an Output command after exiting the program.

However, when the **Output** I/O command is used to display program output, it is wise to have the program end by clearing the Home screen. Clearing the Home screen is necessary in this situation because the calculator may type over the **Output** item when you use it to evaluate an arithmetic expression or execute a command after exiting the program, as in Figure B-10. This figure shows what happens when you use the calculator after executing the program in Figure B-6.

Figure B-10:
The consequence of not clearing the Home screen.

```
NORMAL FLOAT AUTO REAL RADIAN MP
ENTER INTEGER < 20
?10
                                    Done
3+3 67*2
HN INTEGER PLUST FIVE IS E
QUAL TO 15
```

Because you want to give the program user a chance to view any output before clearing the Home screen from a program, place the **Pause** control command before the **ClrHome** I/O command in the program. (The **Pause** control command is discussed in Appendix C.)

Better yet, put the CLRHOME program in the first screen in Figure B-11 on your calculator, and have your program call it whenever you want your program to enable the program user to view the program output before the program clears the Home screen. Calling an external program from within a program is discussed in Appendix C. The second screen in Figure B-11 illustrates what happens when the **prgm** CLRHOME command is placed at the end of a program like that shown in Figure B-6: The program invites the user to press ENTER, and when the user does so, the program clears the Home screen.

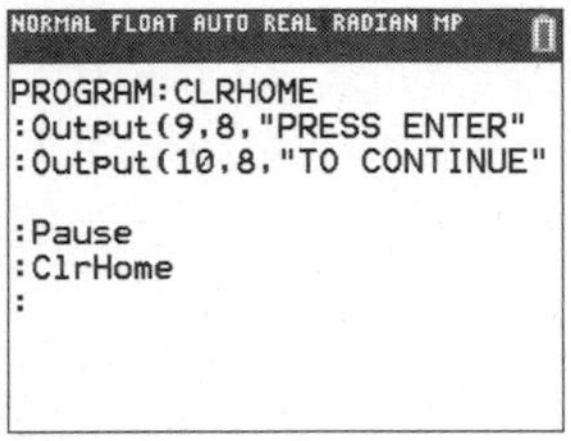

Figure B-11: Using the CLRHOME program to clear the Home screen.

```
NORMAL FLOAT AUTO REAL RADIAN MP

PROGRAM:CLRHOME
:Output(9,8,"PRESS ENTER"
:Output(10,8,"TO CONTINUE"

:Pause
:ClrHome
:
```

ClrHome Program

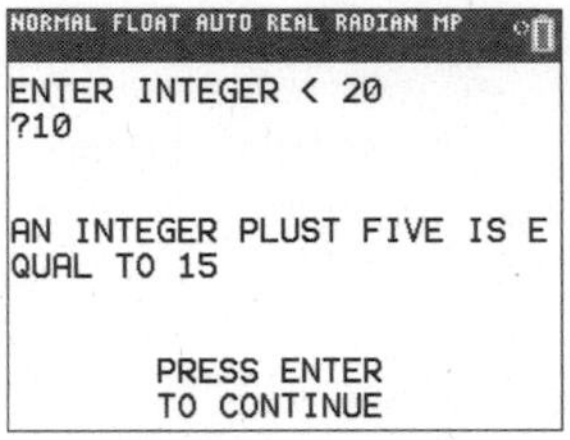

```
NORMAL FLOAT AUTO REAL RADIAN MP

ENTER INTEGER < 20
?10

AN INTEGER PLUST FIVE IS E
QUAL TO 15


        PRESS ENTER
        TO CONTINUE
```

Result

Appendix C

Controlling Program Flow

. . . .

In This Chapter

▶ Entering program control commands in your program

▶ Using decision commands (If, If . . . Then . . . End, If . . . Then . . . Else . . . End)

▶ Using looping commands (While . . . End, Repeat . . . End, For . . . End)

▶ Using branching commands (Goto, Menu)

▶ Stopping the execution of a program

▶ Pausing the execution of a program

▶ Using an external program as a subroutine in your program

. . . .

The flow of a program is controlled by decision commands such as **If . . . Then . . . Else . . . End**, looping commands such as **For . . . End**, and branching commands such as **Goto**. Calling another program from within your program also controls the flow of a program. This chapter explains how to use these and other commands that control the flow of your program.

Entering Control Commands in a Program

The Program Control menu, which houses the control commands, is available only when you're using the Program editor to create a new program or to edit an existing program. A screen of the Program Control menu appears in Figure C-1. (Appendix A explains creating and editing programs.)

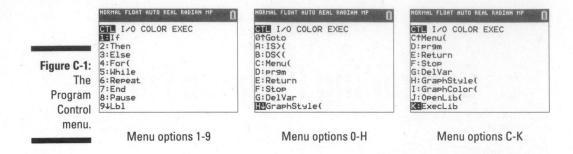

Figure C-1:
The
Program
Control
menu.

Menu options 1-9   Menu options 0-H   Menu options C-K

To enter a control command in a program being written on the calculator, press PRGM, use ▼ to move the indicator to the desired control command, and then press ENTER. The command is then entered at the location of the cursor in the Program editor.

Using Decision Commands

The calculator can handle three decision commands (**If**, **If . . . Then . . . End**, and **If . . . Then . . . Else . . . End**). This section describes how to use them in a program.

The If command

The structure of the **If** command appears in the first screen in Figure C-2. If the condition following the **If** command is true, the program executes the command following the **If** statement (Command 1) and then moves on to the next command in the program (Command 2). If the condition following the **If** command is false, the program skips the command following the **If** statement (Command 1) and then moves on to the next command in the program (Command 2).

An example of using the **If** command appears in the second screen in Figure C-2. The program in this screen gives a 10 percent discount on items that cost $50 or more. The input and output commands (Input, Disp) in this program are housed in the Program I/O menu, which is accessed by pressing PRGM ▶. Commands in this menu are explained in Appendix B. You can enter the inequality that appears in this screen by pressing 2nd MATH 4.

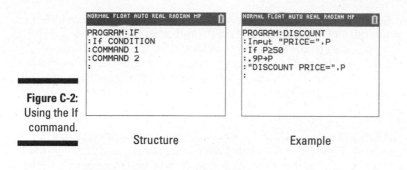

Figure C-2:
Using the If
command.

Structure   Example

The If ... Then ... End Command

The structure of the **If ... Then ... End** command appears in the first screen
in Figure C-3. If the condition following the **If** command is true, the program
executes the commands between **Then** and **End** (Commands 1) and then
moves on to the next command in the program (Command 2). If the condition
following the **If** command is false, the program skips the commands between
Then and **End** (Commands 1) and then continues on to the next command in
the program (Command 2).

An example of using the **If ... Then ... End** command appears in the second
screen in Figure C-3. The program in this screen gives a 10 percent discount
on items that cost $50 or more and then takes off another $10 if the dis-
counted cost is over $100.

Figure C-3:
Using the
If ...
Then ...
End
command.

```
NORMAL FLOAT AUTO REAL RADIAN MP
PROGRAM:IFTHEN
:If CONDITION
:Then
:COMMANDS 1
:End
:COMMAND 2
:
```

```
NORMAL FLOAT AUTO REAL RADIAN MP
PROGRAM:DISPRICE
:If P≥50
:Then
:.9P=P
:If P≥100
:P-10→P
:End
:Disp "DISCOUNTED PRICE=",
P
```

Structure   Example

The If ... Then ... Else ... End Command

The structure of the **If ... Then ... Else ... End** command appears in the
first screen in Figure C-4. If the condition following the **If** command is true,

the program executes the commands between **Then** and **Else** (Commands 1), skips the commands between **Else** and **End** (Commands 2), and then moves on to the next command in the program (Command 3). If the condition following the **If** command is false, the program skips the commands between **Then** and **Else** (Commands 1), executes the commands between **Else** and **End** (Commands 2), and then moves on to the next command in the program (Command 3).

An example of using the **If . . . Then . . . Else . . . End** command appears in the second screen in Figure C-4. The program in this screen divides a number by 2 if it is even, or adds 3 to the number if it isn't. So that the program in the second screen in Figure C-4 fits all on one screen, I used the colon symbol to separate two commands (instead of placing the commands on separate lines.)

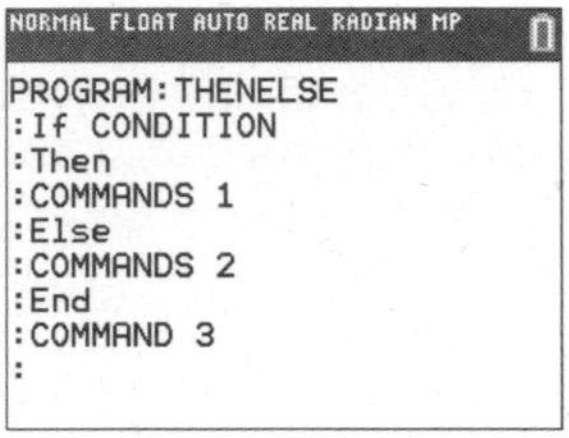

Figure C-4:
Using the
If . . .
Then . . .
Else . . . End
command.

```
NORMAL FLOAT AUTO REAL RADIAN MP
PROGRAM:THENELSE
:If CONDITION
:Then
:COMMANDS 1
:Else
:COMMANDS 2
:End
:COMMAND 3
:
```
Structure

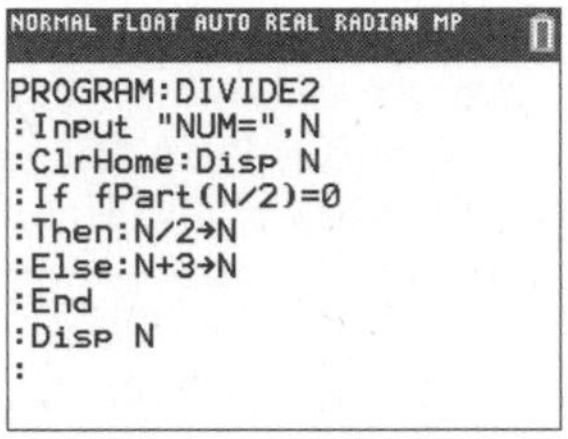

```
NORMAL FLOAT AUTO REAL RADIAN MP
PROGRAM:DIVIDE2
:Input "NUM=",N
:ClrHome:Disp N
:If fPart(N/2)=0
:Then:N/2→N
:Else:N+3→N
:End
:Disp N
:
```
Example

Using Looping Commands

The calculator can handle three looping commands (**While . . . End**, **Repeat . . . End**, and **For . . . End**). This section describes how to use them in a program.

The While . . . End command

The structure of the **While . . . End** command appears in the first screen in Figure C-5. If the condition following the **While** command is true, the program executes the commands between **While** and **End** (Commands 1) and then returns to the **While** command to see whether the condition following it is still true. If it is, the program again executes the commands between **While** and **End** (Commands 1) and then returns to the **While** command to see whether the condition following it is still true. If the condition following the **While** command is false, the program skips the commands between **While** and **End** (commands 1) and then moves on to the next command in the program (Command 2).

To make the **While** command work, the commands appearing between **While** and **End** (Commands 1) must change the value of the variable used in the condition that follows the **While** command. If the value of this variable does not change and the condition is true, you wind up in an *infinite loop*. That is, the calculator continues to execute the **While** command until you stop it or the batteries die.

If you find that your program inadvertently contains an infinite loop (or if it is just taking too long to execute the program and you'd like to stop the execution), press ON. You are then confronted with the ERROR: BREAK error message, which gives you the option to **Quit** the execution of the program.

An example of using the **While . . . End** command appears in the second screen in Figure C-5. The program in this screen starts with the given integer N and divides it by 2 if it is even; if it isn't, it adds 3 to the N. The program then takes the resulting number and divides it by 2 if it is even, or adds 3 to it if it isn't. This process continues until the resulting number is 1. The first **End** command appearing in this program marks the end of the **If . . . Then . . . Else . . . End** command; the second marks the end of the **While . . . End** command.

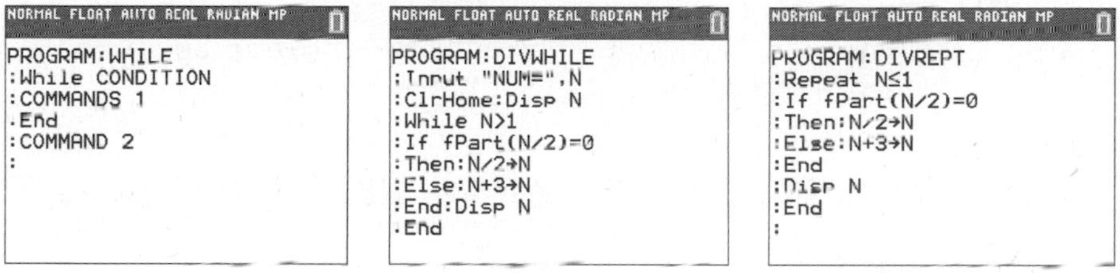

Figure C-5:
Using the
While . . .
End and
Repeat . . .
End
commands.

```
NORMAL FLOAT AUTO REAL RADIAN MP
PROGRAM:WHILE
:While CONDITION
:COMMANDS 1
:End
:COMMAND 2
:
```
Structure

```
NORMAL FLOAT AUTO REAL RADIAN MP
PROGRAM:DIVWHILE
:Input "NUM=",N
:ClrHome:Disp N
:While N>1
:If fPart(N/2)=0
:Then:N/2→N
:Else:N+3→N
:End:Disp N
:End
```
DIVWHILE program

```
NORMAL FLOAT AUTO REAL RADIAN MP
PROGRAM:DIVREPT
:Repeat N≤1
:If fPart(N/2)=0
:Then:N/2→N
:Else:N+3→N
:End
:Disp N
:End
:
```
DIVREPT program

The Repeat . . . End Command

The **While . . . End** and **Repeat . . . End** commands are similar, but opposite. They are similar because they have the same structure (refer to Figure C-5). And they are opposite because the **While . . . End** command executes a block of commands *while* the specified condition is true, whereas the **Repeat . . . End** command executes a block of commands *until* the specified condition is true.

In a **Repeat . . . End** command, if the condition following the **Repeat** command is false, the program executes the commands between **Repeat** and **End** and then returns to the **Repeat** command to see whether the condition following it is still false. If it is, the program will again execute the commands between **Repeat** and **End** and then return to the **Repeat** command to see whether the condition following it is still false. If the condition following the **Repeat** command is true, the program skips the commands between **Repeat** and **End** and then moves on to the next command in the program.

Refer to the third screen in Figure C-5 for an example of using the **Repeat . . . End** command. (The program in this screen is the same one described at the end of the preceding section.)

The For . . . End Command

The structure of the **For . . . End** command appears in the first screen in Figure C-6. When the **For** command is first encountered by your program, it assigns the variable **var** the value in **Start** and then executes the commands appearing between **For** and **End** (Commands 1). It then adds the increment **inc** to the variable **var**. If **var** is less than or equal to the value in **Stop**, the process is repeated. If it isn't, the program moves on by executing the command appearing after **End** (Command 2).

An example of using the **For . . . End** command appears in the second screen of Figure C-6. The results of executing this program appear in the third screen in this figure.

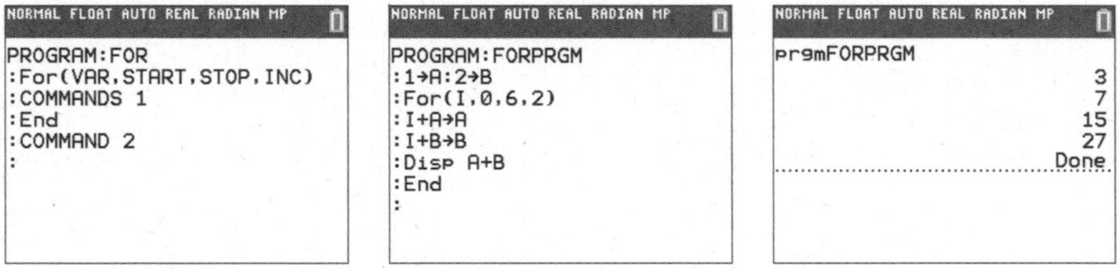

NORMAL FLOAT AUTO REAL RADIAN MP	NORMAL FLOAT AUTO REAL RADIAN MP	NORMAL FLOAT AUTO REAL RADIAN MP
PROGRAM:FOR	PROGRAM:FORPRGM	prgmFORPRGM
:For(VAR,START,STOP,INC)	:1→A:2→B	3
:COMMANDS 1	:For(I,0,6,2)	7
:End	:I+A→A	15
:COMMAND 2	:I+B→B	27
:	:Disp A+B	Done
	:End	
	:	
Structure	Example	Result

Figure C-6: Using the For . . . End command.

Using Branching Commands

The calculator can handle two branching commands: **Goto** and **Menu**. This section describes how to use them in a program.

Using the Goto command

The **Goto** command is used in conjunction with the **Lbl** (Label) command. The **Goto** command sends the program to the corresponding **Lbl** command. The program then executes the commands that follow the **Lbl** command. To ensure that the program knows which label (**Lbl**) to go to, be sure to give the label a one- or two-character name that consists of letters, numbers, or the Greek letter θ. The **Goto** command then refers to this name when telling the program which label (**Lbl**) to go to, as shown in Figures C-7 and C-8. The **Goto** command directs the program to a subroutine contained in the program, or terminates the program when a specified condition is satisfied. These situations are explained in the remainder of this section.

The structure for using the **Goto** command to direct the program to a subroutine contained in the program appears in the first screen in Figure C-7. The subroutine consists of the commands that are designated by Commands 2 in this screen. The program in this screen executes Commands 1, executes Commands 2, and then (if the condition following the **If** command is true), it executes Commands 2 again. It continues to re-execute Commands 2 until the condition following the **If** command is false. Then it continues with the program by executing Commands 3.

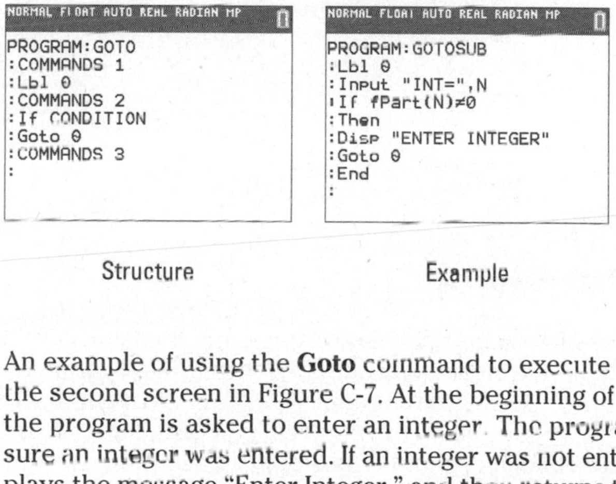

Figure C-7:
Using the Goto command to execute a subroutine.

```
NORMAL FLOAT AUTO REAL RADIAN MP

PROGRAM:GOTO
:COMMANDS 1
:Lbl θ
:COMMANDS 2
:If CONDITION
:Goto θ
:COMMANDS 3
:
```

Structure

```
NORMAL FLOAT AUTO REAL RADIAN MP

PROGRAM:GOTOSUB
:Lbl θ
:Input "INT=",N
:If fPart(N)≠0
:Then
:Disp "ENTER INTEGER"
:Goto θ
:End
```

Example

An example of using the **Goto** command to execute a subroutine appears in the second screen in Figure C-7. At the beginning of the program, the user of the program is asked to enter an integer. The program then checks to make sure an integer was entered. If an integer was not entered, the program displays the message "Enter Integer," and then returns the user to the beginning of the program, once again asking the user to enter an integer. If an integer is entered, the program continues with the commands that come after the **If . . . Then . . . End** command appearing in this screen. The request to have the user enter an integer constitutes the subroutine in this program.

When the **Goto** command directs a program to a label (**Lbl**), that label can appear in the program either before or after the **Goto** command. If it appears after the **Goto** command, the program skips executing all commands that are between the **Goto** command and the corresponding **Lbl** command.

The structure for using the **Goto** command to terminate a program appears in the first screen in Figure C-8. In this theoretical program, the program executes Commands 1, and then it continually executes Commands 2 until the condition after the **If** command is false. The program is terminated by the **Stop** command only when the condition appearing after the **If** command is false.

An example of a program that uses the **Goto** command to terminate a program appears in the second screen in Figure C-8. The program in this screen asks the user to enter a number. If the number is less than 1,000, the program displays the square of that number and then prompts the user for another number. The program continues in this fashion until the user enters a number that is greater than or equal to 1,000.

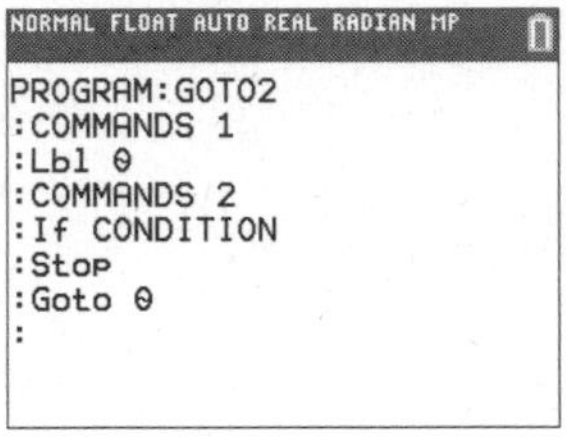

Figure C-8:
Using the Goto command to terminate a program.

```
NORMAL FLOAT AUTO REAL RADIAN MP
PROGRAM:GOTO2
:COMMANDS 1
:Lbl θ
:COMMANDS 2
:If CONDITION
:Stop
:Goto θ
:
```

Structure

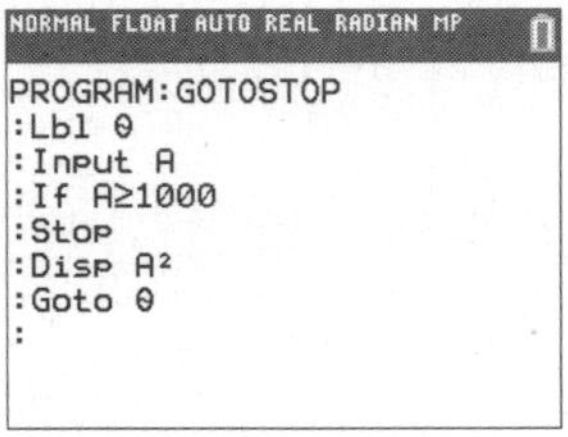

```
NORMAL FLOAT AUTO REAL RADIAN MP
PROGRAM:GOTOSTOP
:Lbl θ
:Input A
:If A≥1000
:Stop
:Disp A²
:Goto θ
:
```

Example

Creating a menu

The **Menu** command is a glorified **Goto** command. It enables the program user to select an item from a menu, and then have the program execute the commands that are specific to that item. After executing the commands that are specific to the chosen item, the program can terminate, return to the menu so the user can make another selection, or continue by executing the commands in the program that appear after the commands that are specific to the chosen menu item.

The first screen in Figure C-9 illustrates the structure of a menu-driven program that terminates after executing the commands associated with the chosen menu item. If, for example, the user of this theoretical program selects ITEM A from the menu, Commands 1 are executed, and then the **Stop** command terminates the program. If the user selects QUIT from the menu, the program clears the Home screen and then terminates because it has no more commands to execute.

It is OK to leave off the right parenthesis at the end of a command in a program, because the program will take up less RAM on the calculator.

The second screen in Figure C-9 illustrates the menu that the user of the program sees. The moving busy indicator in the upper-right corner is the calculator's way of telling the user that it is waiting for a menu item to be selected.

When you create a menu-driven program, it's a common courtesy to offer QUIT as a menu item. This enables the user to quickly exit the program if he or she inadvertently selects the wrong program to execute.

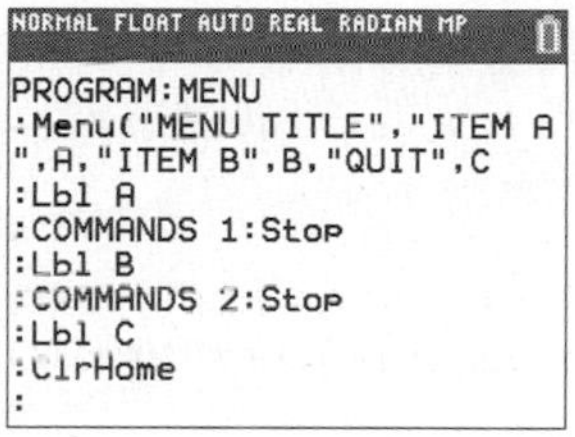

Figure C-9:
A terminating menu-driven program.

```
NORMAL FLOAT AUTO REAL RADIAN MP
PROGRAM:MENU
:Menu("MENU TITLE","ITEM A
",A,"ITEM B",B,"QUIT",C
:Lbl A
:COMMANDS 1:Stop
:Lbl B
:COMMANDS 2:Stop
:Lbl C
:ClrHome
:
```

Structure

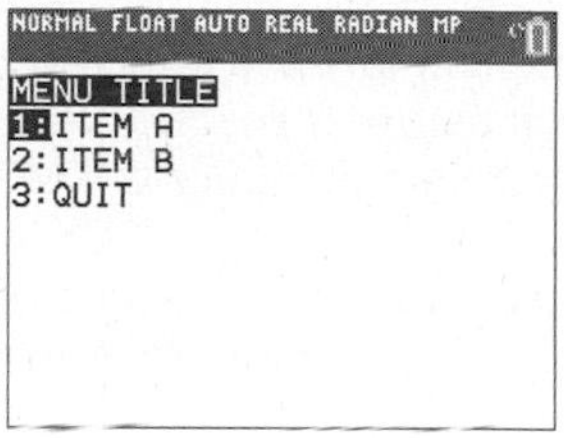

```
NORMAL FLOAT AUTO REAL RADIAN MP
MENU TITLE
1:ITEM A
2:ITEM B
3:QUIT
```

Menu

The first screen in Figure C-10 illustrates the structure of a menu-driven program that returns the user to the menu after he has selected and executed a menu item. If, for example, the user of this program selects THIS from the menu, the calculator executes the commands housed in the external program named THIS, and then returns the user to the menu to make another selection. The external program named THIS is pictured, in its entirety, in the second screen in Figure C-10. If the user selects QUIT from the menu, the program clears the Home screen and terminates because there are no more commands in the program for it to execute.

When you create a menu-driven program that repeatedly returns the program user to the menu, it's wise to supply the program with a means of terminating itself. Adding a QUIT option to the menu is an easy way to do so.

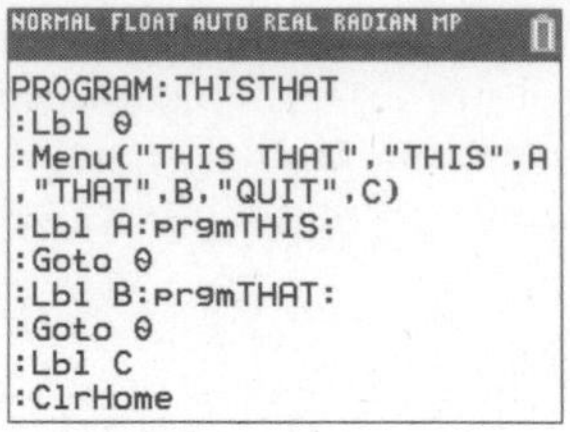

Figure C-10:
A menu-driven program that returns the user to the menu.

```
NORMAL FLOAT AUTO REAL RADIAN MP
PROGRAM:THISTHAT
:Lbl θ
:Menu("THIS THAT","THIS",A
,"THAT",B,"QUIT",C)
:Lbl A:prgmTHIS:
:Goto θ
:Lbl B:prgmTHAT:
:Goto θ
:Lbl C
:ClrHome
```

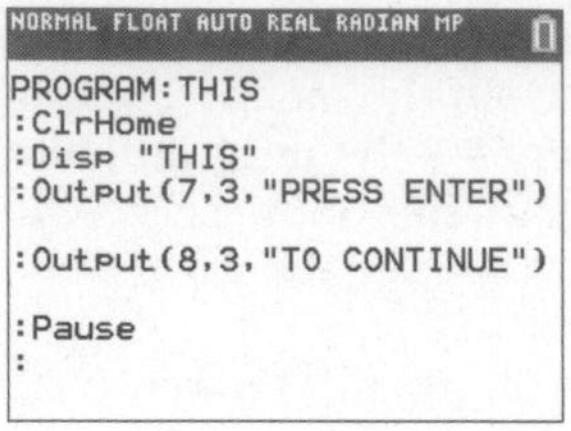

```
NORMAL FLOAT AUTO REAL RADIAN MP
PROGRAM:THIS
:ClrHome
:Disp "THIS"
:Output(7,3,"PRESS ENTER")

:Output(8,3,"TO CONTINUE")

:Pause
:
```

Structure   Called Program

Stopping a Program

To stop a program while it is executing, press [ON]. You are then confronted with the ERROR: BREAK error message that gives you the option to QUIT the execution of the program.

The control command **Stop** is added to a program when you want to terminate the program before it reaches the end. It is illustrated in Figure C-8 and in the first two screens of Figure C-9.

Placing the **Stop** command at the end of a program isn't necessary. The program automatically terminates execution when it reaches the last command.

Pausing a Program

When a program is executed, the output from the program is displayed quickly on the Home screen or in a graphing window. Sometimes it is necessary to pause the program so that the program user has time to view the results of a program output.

The **Pause** command temporarily suspends the execution of a program so that the user can see the program output. The execution of the program is resumed when the program user presses [ENTER], as in the program in the second screen in Figure C-10. The program output appears in Figure C-11. The moving busy indicator in the upper-right corner of Figure C-11 tells the program user that the program is waiting for the user to press [ENTER] to resume execution of the program.

Because most program users don't realize that they must press [ENTER] to resume the execution of a paused program, I like to precede the **Pause** command in the program with the reminder that the user must "press enter to continue," as illustrated in the second screen in Figure C-10. The consequence of doing this appears in Figure C-11.

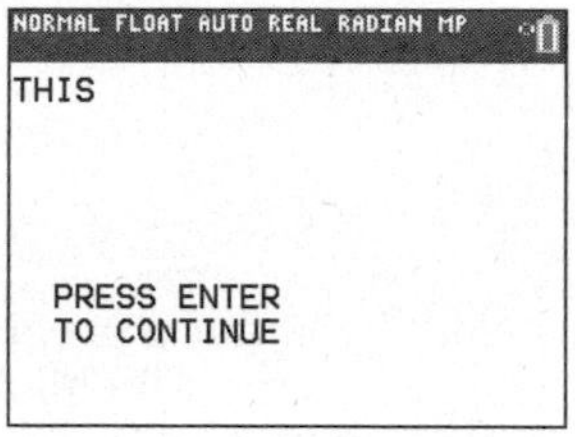

Figure C-11:
A paused
program.

```
NORMAL FLOAT AUTO REAL RADIAN MP
THIS



PRESS ENTER
TO CONTINUE
```

Executing an External Program as a Subroutine

It's quite easy to have a program call and execute another program saved on your calculator, and then return to the original program to complete its execution of that program. One command accomplishes the processes of calling, executing, and returning: the **prgm** command (accessed by pressing [PRGM][ALPHA][x^{-1}]). The name of the program being called is placed directly after the command, as shown in the two screens in Figure C-10. Notice that there is no space between the command **prgm** and the name of the program.

After the externally called program is executed, the calling program continues to execute the commands that follow the **prgm** command *provided that* the externally called program does not encounter the **Stop** command. This command terminates both the called and calling programs. As an example, if the program GOTOSTOP in the second screen in Figure C-8 is called by your program, then when the program user enters a number greater than or equal to 1,000, both the calling and called programs terminate.

If you want the externally called program to return control to the calling program *before* it completes its execution, you do so by putting the **Return** command in the appropriate place in the externally called program. As an example, consider the program GOTORTRN appearing in Figure C-12. GOTORTRN is simply the program GOTOSTOP (second screen in Figure C-8) with the **Stop** command replace by the **Return** command. If your program calls GOTORTRN, then when the program user enters a number greater than or equal to 1,000, the GOTORTRN program is terminated and the calling program continues to execute.

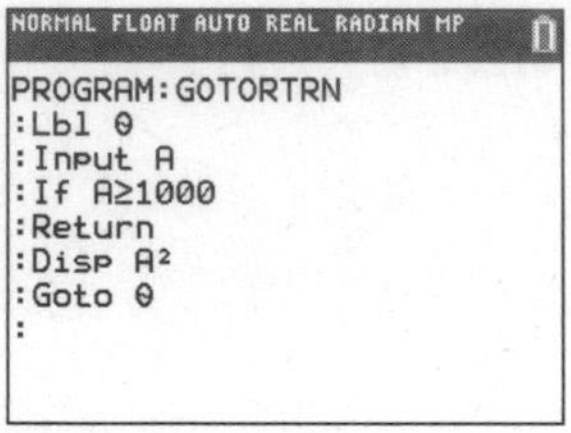

Figure C-12:
Using the
Return
command
in a called
program.

```
NORMAL FLOAT AUTO REAL RADIAN MP
PROGRAM:GOTORTRN
:Lbl 0
:Input A
:If A≥1000
:Return
:Disp A²
:Goto 0
:
```

If a program containing a **Stop** command is called by another program, that command may terminate the execution of *both* programs. If the **Stop** commands in the called program are replaced with the **Return** command, then after the called program is executed, program control returns to the calling program.

Index

• 𝑇 •